Candle MAGIC

Copyright © MMII by De Agostini UK Limited. All rights reserved. No part of this publication may be totally or partially reproduced, stored in a retrieval system or transmitted in any form by any means electronic, mechanical, photocopying recording, or otherwise, without first obtaining the publisher's prior written authorization.

Published in the United States by
TODTRI Book Publishers
254 West 31st Street
New York, NY 10001-2813
Fax: (212) 695-6984
E-mail: info@todtri.com

Visit us on the web!
www.todtri.com

ISBN 1-57717-206-X

Packaged by De Agostini Rights/mb.dh.sm

Cover design by WDA

Printed and bound in Korea

cont

introduction

- 6   the candle code
- 9   intro

decoration candles

- 12   coloured-layered cones
- 16   rolled beeswax candles
- 20   lavender aromatherapy candles
- 24   ice blue
- 28   floating candles

making candles

- 34   gilded leaves
- 38   warm hearts
- 40   applique wax motifs
- 44   carving dipped candles
- 46   glitter tea lights

holders lanterns & containers

50	shine a light
54	orange pomander holders
56	salt-dough candle holders
60	mosaic votive
64	chinese paper lanterns

creating displays & gifts

68	radiant blooms
72	giftwrap candle cones
76	pastel glow
78	black magic
80	metal-foil candle collars

inspirations

86	ideas from antiquity
88	reflective glow
90	bathroom bliss
92	candles for a summer evening

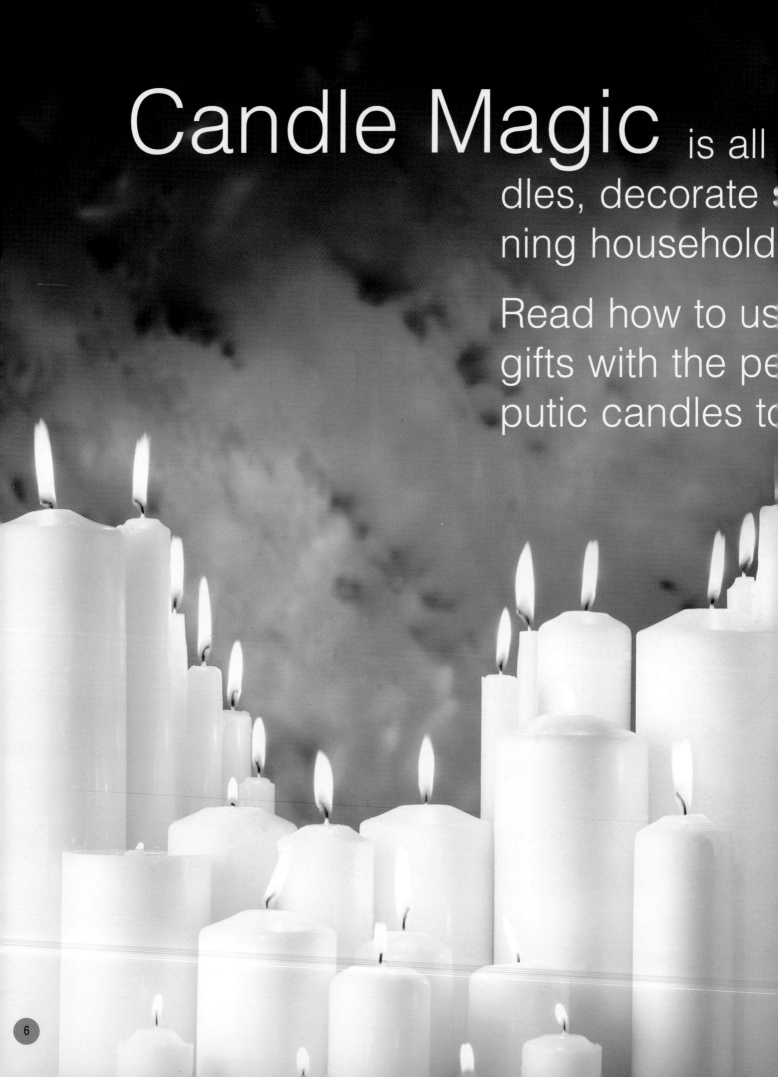

Candle Magic
is all
dles, decorate s
ning household

Read how to us
gifts with the pe
putic candles to

u need to make exclusive designer can-
re-bought candles to put together stun-
ecorations and seasonal displays.

candles to brighten up your home, make
onal touch, or use scented and thera-
reate a pleasant atmosphere.

Candles and candlemaking are perfectly safe,

When using candles...

Always follow these basic rules:

Always use the right type of candle holder for the candle.
Never leave a burning candle unattended even for a short time.
Always keep candles away from children.
Leave at least 5cm between candles when several are burning together.
Don't move a candle while it is lit.
Don't light an outdoor candle indoors.

POSITIONING CANDLES
Only use candles on heatproof surfaces, away from draughts and not near curtains, soft furnishings or other flammable materials. Never place a candle above a heat source such as a radiator or even a television, as this can cause the wax to melt or the candle to bend.

HOLDERS
Stability is essential. Narrow candles should be in a holder that fits the base tightly. Wide candles should be used with a spiked candleholder. Very wide candles (over 7.5cm) can be used on a flat, heatproof dish, so long as they are stable and unlikely to fall over. Church candles and the like are designed to burn down completely, so the holder must be able to contain any melted wax that leaks out.

NIGHTLIGHTS AND TEALIGHTS
Although tealights and nightlights look the same, they are made for different uses. Tealights were mainly designed for warming food. They burn for about five hours. Nightlights were designed to provide a low level light for up to eight hours. Their different burning characteristics mean that if you put a tealight in a nightlight holder, the holder may overheat, causing it to melt or shatter.

WAX BUILD-UP
Don't allow too much melted wax to build up, especially with wide candles. It can be dangerous if the wax suddenly leaks away. Such candles are really designed only to be burned for around 3-4 hours at a time. Remove excess wax build-up by cutting or pouring it away after putting out the candle.

WICKS
Keep the wick of a normal indoor candle trimmed to under 1cm to avoid a high flame, which can cause the candle to smoke or flare. Pieces of burnt wick should be removed from the pool of wax, as they can catch fire from the candle flame and burn.

PUTTING OUT CANDLES
Try to avoid blowing out a burning candle as this will cause it to smoke and also risks spreading molten wax – even if you shield it with a cupped hand. The traditional method for putting out a candle is to use either a conical or scissors-type snuffer, but these can't be used on very wide candles. The best method here is simply to use the point of a knife or a flameproof stick to push the wick into the pool of wax to drown it, then lift it back up again.

...ong as you observe some simple precautions.

When making candles...

Treat candlemaking with similar precautions to cookery. For example, handle melted wax with the same care that you would when working with hot cooking oil. Always follow any special safety notes provided with the candlemaking method you are following.

HEATING WAX
Wax is perfectly safe at the temperature required to melt and cast it. But whenever you melt wax, it is important not to overheat it: if the wax gets too hot and vapourises, it may become a fire hazard.

- Melting wax in a double boiler or pan of water will prevent overheating, as it cannot get hotter than 100°C, the boiling point of water. Even so, never leave the pan unattended and ensure that it does not boil dry.
- If you have a wax thermometer, you can use this to monitor the melting process, keeping the wax at 85°C, the ideal working temperature.
- Be careful whenever you remove the wax from the heat, as any spills could catch fire from the burner. To prevent this, turn off the heat before you remove the wax container. Handle the wax container with the same care that you would use for a container of freshly boiled water.

FIRE PRECAUTIONS
Overheated wax will begin to smoke and produce acrid fumes. If this happens turn off the heat and allow the wax to cool. If the worst happens when heating wax:

- Do not attempt to move the pan.
- Never try to douse the fire with water – it will only spread the flames.
- Turn off the heat.
- Smother the flames with a pan lid, fire blanket or a damp cloth.

CLEANLINESS
To avoid messy spills, always:

- Wear old clothes and cover working surfaces with newspaper or a board.
- Give yourself plenty of room and do not work near rugs or carpets.
- Keep all the tools and materials near to hand.
- Have some sheets of old newspaper to hand to mop up spills.

TO CLEAN UP SPILT WAX:

- On carpets and cloth, scrape off the excess and remove the rest by pressing with a hot iron through a pad of kitchen paper to absorb the melted wax. Repeat with fresh paper until no more wax is present.
- On metal or plastic objects, either place in boiling water to melt the wax and float it away, or place in the freezer to make the wax brittle so that pieces can easily be flaked off.

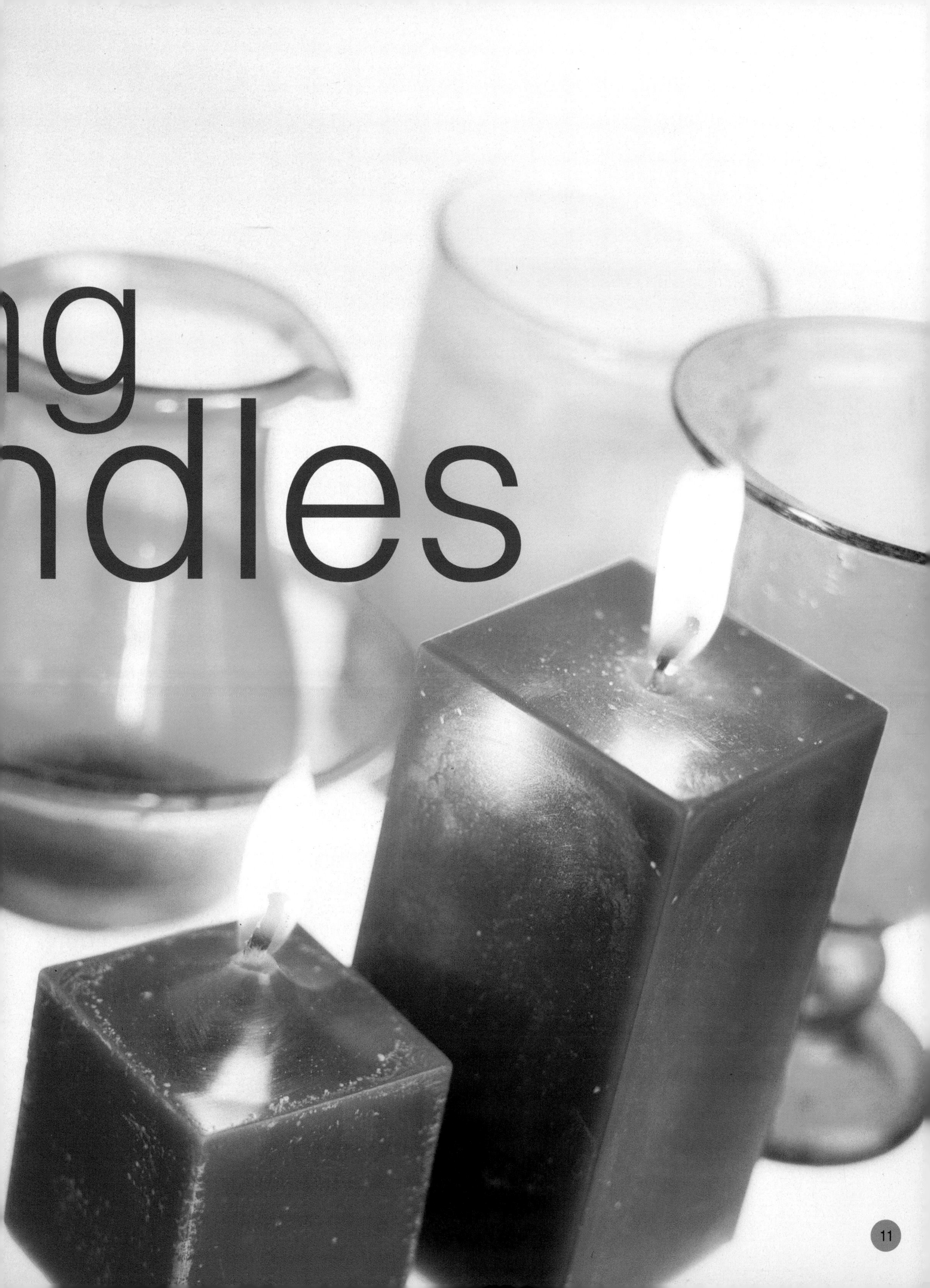

ng
ndles

colour-layered cones

It's possible to make unique designer candles with a minimum of special equipment. Here's a way to make subtly tinted cones using household items and cheap shop-bought candles as a source of wax.

Candles can be cast very easily and successfully using a variety of everyday items as moulds. This project uses an ordinary plastic kitchen funnel to create a stylish conical candle with layers of colour.

To suit the beginner, we show a method that doesn't use any special tools or materials. The coloured wax is obtained simply by melting down candles of the right colour. However, if you already have some candlemaking equipment, including supplies of candlemaking wax, additives and dyes, the step-by-step instructions explain where you can use them. We also explain how you can obtain different colour effects within the stripes, as seen in the main photograph opposite, simply by allowing the melted wax to cool in the funnel for a longer or shorter time.

Shop-bought candles
Cheap, ready-made candles provide an ideal source of small amounts of coloured wax, and you can often recycle the wicks, too.

1 ▶ MELT THE WAX

If you don't have a candlemaker's double boiler, melt your wax in cleaned food tins standing in a shallow pan of boiling water. To keep the tins upright while the wax melts, you can pack any spaces in the pan with clean stones or

TIP If you are only making a single candle, you only need one tin. Use it to melt the white wax, then the yellow, and finally the orange. However, you will need to work quickly if you want to prevent the first layers from hardening while you melt the wax for the next ones. Using a different can for each colour makes it easier to control the process, and is essential if you want to make more than one candle at once (using more than one funnel).

You will need:

- **Either: white candles, pastel yellow candles and pastel orange candles Or: white candles, plus small bright yellow and red candles Or: candlemaker's paraffin wax granules, stearin, dye discs** (see Working with Coloured Wax, overleaf)
- **Wick** (recycle the existing wicks if you are using ready-made candles)
- **Blu-Tack or candlemaker's mould seal**
- **Plastic kitchen funnel** There are various diameters that you can use to create different sizes of candles. It is possible to cast a small candle in a large funnel, by not filling the cone right up with wax, although this will leave you with a larger spout to plug.
- **Cleaned tin cans and an inexpensive shallow pan, Or: a purpose-made double boiler**
- **Barbecue skewers**
- **Candlemaker's thermometer** (not essential).

How much wax do you need?
The amount of wax depends on the diameter of your cone. A 10cm cone, striped as shown, needs roughly 25g of white wax, 45g of yellow wax and 130g of orange wax. A candle twice the diameter needs eight times (not twice) as much. When weighing out the candle wax, allow a little over to avoid running short.

2 ▶ PREPARE THE WICK

When the candles have melted, fish out the wicks with a wooden skewer. Pull them out straight and put them on one side to harden. Take a blob of Blu-Tack (or candlemaker's mould seal) large enough to wedge into the

3 ◀ **PREPARE THE MOULD** Stand your funnel in a heavy glass jar to keep it upright during the moulding process. You shouldn't get any leakage from the spout, but even if you do, the jar will catch it. Stand the jar on a heatproof surface from which you will easily be able to clean any drips of wax. Push the Blu-Tack end of the wick firmly into the funnel and press it down flush with the end of the neck. To keep the wick upright during the pouring process, take a barbecue skewer and spike it through the wick so that the skewer will lie across the top of the funnel **(inset)**.

SAFETY!

• Wax is perfectly safe at the temperature required to melt and cast it. But whenever you melt wax it is important not to overheat it: if the wax gets too hot and vapourises, it may become a fire hazard. Melting it in a pan of water will prevent this, as the wax cannot get hotter than 100°C, the boiling point of water. Even so, never leave the pan unattended and ensure that it does not boil dry.

4 ▼ **POUR THE FIRST LAYER** Carefully remove the can of white wax from the heat and pour it into the funnel to a depth of roughly one-third of the funnel. Pour it in the centre to avoid getting drips down the sides of the funnel. It will start hardening at once, but it will take several minutes until the wax is solid enough to pour the next layer. You can check how the process is going on by carefully pressing on the wax near the centre, where it takes longest to set.

5 ▲ **POUR THE SECOND LAYER** The look of your candle can be altered by varying how long you let the layers harden. If you pour a new layer while the first is soft (after about 10 minutes, say), the colours will blend to give a subtle graduation. If you let the first layer harden completely, the second will form a distinct stripe. Avoid pouring before the first layer can support the second, or the colours will merge totally. Pour in the yellow wax to a depth of around two thirds of the funnel.

Options...

6 ▶ POUR THE THIRD LAYER When the yellow wax is firm, complete the moulding process by pouring a layer of orange wax up to the top of the tapered part of the funnel. Allow the wax to harden and cool thoroughly before attempting to remove your finished candle.

7 ▼ RELEASING THE CANDLE Release the candle from the mould by using a chopstick or pencil to push up the neck of the funnel. Don't try pulling the wick as this won't be strong enough. When you have removed the candle, remove any Blu-Tack that is clinging to the wick. You can also clean up the point of the cone by using a knife to scrape it smooth. Cut off the surplus wick from the base of the cone and trim the wick projecting from the top to a length of about 1.5cm.

Working with coloured wax

The easy way to get a small amount of an exact wax colour is to melt down some candles that are the shade you want. The simplest way to complete this project is to use white candles, yellow candles and orange candles (or different colours of your choice) to create the layers.

However, you can produce any shade you want by blending colours with plain wax. If you are working with candlemaker's clear wax granules and powdered stearin, the normal method is to blend in some wax dye (commonly supplied as wax tablets containing concentrated colours).

An alternative way of creating subtle shades when you don't have the appropriate wax dye is simply to add pieces of coloured candle to the mix. For example, you can create a pastel orange by adding small amounts of yellow wax, plus a little red, to a large amount of white wax.

Creating wax colours is rather more of an art than a science. This is partly because the amount of wax dye needed depends on the quantity of plain wax you have – and it isn't always easy to measure either the wax or the dye exactly. The other reason is that when wax is molten, you can't predict the exact colour that it will appear when cold.

As a quick check on the shade, pour a little melted wax on to a clean white plate (or a piece of greaseproof paper). It will harden quickly on contact with the cold china, revealing its true colour. You can use this method to adjust the colour, adding more coloured wax to darken the mix or alter its shade, testing the wax again at each stage.

No problem!

You shouldn't normally have any difficulty removing the cone from the funnel, as its shape makes it easy to release. If it does stick, however, stand the funnel upright in hot water for a few moments to soften the surface of the wax that is in contact with the plastic, then try again.

rolled beeswax candles

Beeswax candles are probably the easiest of all candles to make, and they smell wonderful when they burn.

Beeswax is sold as flat sheets, which you simply roll into a tube so that they enclose the wick.

Fragrant natural beeswax made by honeybees was used for the first candles in history and is still much prized today. It burns with a bright clear flame and a delicious honey scent. For candlemaking, it is normally supplied in the form of the thin sheets used here, which are embossed with a honeycomb pattern. These are simply rolled around a length of wick so that they melt together as the wick burns down. We have used two sheets of natural beeswax measuring approximately 15cm x 10cm that can be rolled lengthways to produce two 15cm candles, each of which will burn for around an hour and a half.

You will need:
- **Natural beeswax sheets (Approx. 15cm x 10cm)**
- **Wick**
- **Scissors**
- **Straightedge or length of wood (optional)**
- **Sharp knife (optional)**

1 ◄ ADD THE WICK Lay out the wax and gently smooth it flat. Cut the wick into two equal lengths. Then press one length down the long edge of one piece of wax. Ensure one end of the wick is aligned. This edge will form the base of your candle.

2 ► ANCHOR THE WICK Turn over the edge of the sheet so that it traps the wick, pressing it tightly into place, as it is the wax that is in contact with the wick which will melt first. It does not matter if you flatten the honeycomb cells on the sheet along this edge.

3 ▼ ROLL THE CANDLE Roll up the candle, taking care that the edge that will become the base stays in line. As the roll builds up, press down the layers gently but firmly. It doesn't matter if there are a few small gaps between the layers as the wax will melt into a pool around the burning wick.

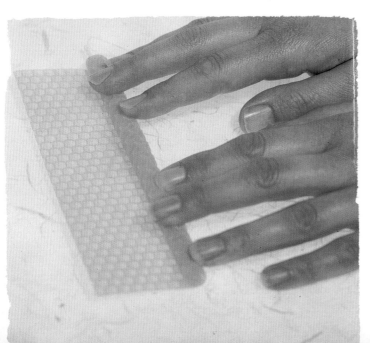

4 ► COMPLETE THE ROLL As you reach the end of the roll, press the edge of the wax down firmly, but take care not to flatten the honeycomb pattern. The wax will stay in place with no tendency to unroll, and your candle is now complete.

Tapered beeswax candles

Instead of rolling straight cylindrical candles, you can use this method to make tapered candles with a spiral effect on the outside. The beeswax is cut on a staggered diagonal to produce two identical halves. As a result, you can create two slightly smaller candles from a single sheet of wax.

1 ▲ CUTTING THE WAX Measure 2cm along from the edge of one short side of the beeswax sheet, and 2cm along, on the opposite side. Lay a straight edge or piece of wood across the sheet, joining up the two points, and cut the wax with a sharp knife or craft knife.

2 ▼ ADDING THE WICK Cut a length of wick at least 5cm longer than the longer of the short sides of the sheet. Lay the wick along the edge of the wax, 2-3mm in, overhanging each end by about 2.5cm.

TIP Beeswax sheets are naturally pliable and roll easily, so don't worry about cracking them. However, when working with narrow sections or in cold conditions, you may find it easier to work with if you warm the beeswax slightly by laying the sheet on a warm – not hot – baking tray.

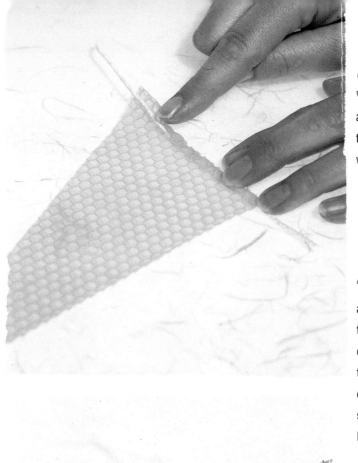

3 ◀ **TRAP THE WICK** Gently pinch the wax up and around the wick along its length to anchor it firmly in position for when you start to roll.

4 ▶ **ROLLING THE CANDLE** Carefully roll up the candle along the length of wax, keeping the base as square to the straight edge as possible, and making the roll quite tight. The cut edge of the wax will spiral along its length.

5 ◀ **TRIMMING THE WICK** Trim the wick level with the base of the wax. Then use your thumbs to gently push the layers of wax around the base into alignment, so that the candle will stand straight and firm.

6 ▶ **COMPLETING THE CANDLE** Trim the wick at the top of the candle to 1cm. The pointed end will burn down much quicker than the thicker part of the candle.

lavender aromatherapy candles

As they burn, these delicately tinted candles give off the mellow, soothing scent of lavender. The key ingredient is a few drops of special candle perfume.

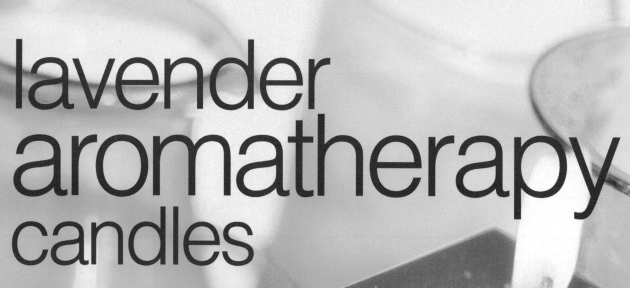

All you need to do to make a scented candle is to add a suitable fragrance to the melted wax before casting it. Candle perfume – here, lavender perfume – is specially formulated for use in candlemaking, and as you only need a few drops each time, it can be used over and over again. The candle itself is a straightforward moulding using melted wax. The method shown uses a square mould improvised from cardboard; if you prefer, you can use a ready-made plastic mould. In the step-by-step instructions we show a method based on using wax pellets and stearin powder, but if you only want to make one or two candles, you can simply melt down ordinary white household candles.

Lavender perfume
Lavender scent is renowned for its healing and beneficial effects. When used in aromatherapy, it promotes feelings of tranquillity.

Purple dye disc
This special wax dye is ideal for making candles.

1 ▶ CUT THE CARD Cut a piece of cardboard 20cm x 7cm. You need to use a rigid type of card such as the back of a notepad. Divide the 20cm length into four equal parts by scoring vertically along the cardboard at 5cm intervals. Use a metal ruler and a craft knife, taking care not to cut right through the card.

2 ▶ ▼ FORM THE MOULD Bend the card along the scored lines, with the cut side outwards. Form it into a square mould, taping the two open edges together with a piece of masking tape. Push the seam tightly together to prevent the wax leaking out. Trim the tape level with the top and bottom of the card, then strengthen the mould by applying strips of masking tape around all four sides.

You will need:
- Piece of stiff cardboard 20cm x 7cm
- Ruler
- Craft knife
- Masking tape
- Plasticine or mould seal
- Disposable plastic lid
- Cocktail stick
- Approximately 15cm of 38mm wick
- Stearin powder
- Wax pellets
- Dye disc
- Lavender scent
- Water spray
- Double boiler or two pans, one old
- Candlemaking thermometer (optional)

3 ▶ MAKE A BASE Find a base large enough to stand the mould on, such as the lid from a disposable cup. Pierce the centre with a cocktail stick or the point of a metal skewer to allow you to insert the wick.

4 ▼ **SEAL THE MOULD** Set the mould on the plastic lid and use plasticine or mould seal to seal the base to the plastic lid. Be careful not to push the square out of shape.

TIP If you have used a see-through plastic lid, you can look through it from underneath to check that the plasticine has sealed all around the edge of the cardboard.

5 ▼ **ADD THE WICK** Push the wick through the hole in the base up into the mould. Leave 2cm of wick protruding through the base and seal the hole with plasticine.

6 ▶ **FIX THE WICK** Pull the wick taut and thread a cocktail stick through the free end, resting it on the sides of the cardboard mould. Make sure the wick is central in the mould.

7 ▶ **MELT THE STEARIN** To work out how much wax and stearin to use, fill the mould with wax pellets, up to height of the finished candle. Tip the wax into a container and then repeat the process. This will give enough wax to make one candle and top up the base. To work out how much stearin to use, weigh the wax and use ten per cent of its weight in stearin. Melt the stearin in an old pan over a pan of hot water, or in a double boiler. Add about one-sixteenth of the dye disc to the melted stearin to give a lavender colour to the candle.

TIP To keep the top of the mould square, place the cocktail stick supporting the wick diagonally across the mould and tape the ends in place.

8 ▶ **MELT THE WAX** Add the wax to the pan and melt it together with the stearin, stirring well with an old metal spoon. If you have a candlemaking thermometer, you should heat the wax to 82°C. To suit the perfume, this is a little cooler than the temperature that is normally used when candlemaking.

9 ▼ ADD THE SCENT Add 2-3 drops from the lavender perfume bottle to the melted wax. You should not allow the temperature to rise above 82º C as it may spoil the scent, so if you don't have a candlemaking thermometer, add the lavender scent as soon as the wax has properly melted.

10 ▶ PREPARE THE MOULD Spray the inside of the mould thoroughly with water or silicon lubricant. This will help prevent the wax sticking to the mould and make the mould easier to remove.

11 ▶ POUR THE WAX Pour the hot wax slowly and carefully into the mould – transfer it to a jug if you don't have a pouring lip on your pan. Fill to the height required (not the top of the mould), keeping a little wax back to top up the candle, as it will shrink a little when the wax hardens. Do not disturb the mould until the wax has set.

12 ▶ TOP UP Before the wax has fully set, reheat the remaining wax and use it to top up the dip that will have formed around the wick at the top of the mould (which will be the base of the candle). Leave the wax to set hard.

Options...
Size & shape

You can easily vary the size and shape of the mould you use. For example, you can make a six-sided candle by cutting the cardboard 30cm wide and dividing it into six 5cm sections. Or you can make a cylindrical candle by using the core from a roll of paper tissue. In all cases, use plenty of masking tape binding to hold the mould in shape and prevent it from bulging or splitting.

13 ▶ REMOVE THE MOULD It will take some time for the candle to cool completely. When the wax is cold, remove the plasticine from around the wick and gently peel away the base from the mould. Remove the masking tape and and peel the cardboard away from the candle. Use water to soak any cardboard sticking to the candle, and polish with a soft cloth. Trim the wick level with the base and cut it to 2.5cm at the top.

ice blue

Ice candles are made by casting molten wax in a mould filled with crushed ice. The holes left when the ice melts are completely unpredictable, creating fascinating free-form shapes – although the candle will burn in the normal way.

The secret of casting successful ice candles is to mould the wax around a narrow candle or taper. This will give the candle a basis from which to burn, melting the tracery of wax around the flame as the wick burns down. Without the solid core provided by the taper, you cannot be sure that the candle will keep burning.

You'll also need a suitable candlemaking mould. We used a square one, which gives a dramatic effect when holes appear through the corners, but a round mould will work just as well. If you don't have any special candlemaking materials and equipment, you can improvise the mould and other items that you need.

Wax and stearin powder
The wax is provided in the form of pellets, making it easy to measure and melt. Based on the weight of wax, allow ten per cent of stearin, which is used to control the melting and burning properties of the wax.

Wax dye
This normally comes in the form of wax discs or sticks impregnated with concentrated dye. You only need a tiny quantity of deep blue dye to produce the ice blue effect.

Crushed ice
There's no need to obtain crushed ice specially. Just break up a few ice cubes from the fridge.

1 ▼ PREPARE THE CORE CANDLE Use a sharp knife or craft knife to cut the narrow candle or taper to the height of the mould, plus 5cm. Then trim away the extra 5cm of wax, taking care not to damage the wick, but leaving it protruding so you can thread it through the mould.

You will need:
- Candle mould – a square one is preferable (see Tip)
- Narrow candle or taper candle
- Mould seal, Blu-Tack or plasticine
- 150gm wax granules (for one 5cm square candle)
- 15gm stearin powder
- Blue wax dye
- Wax thermometer
- Ice cubes
- Sharp knife or craft knife
- Plastic bag
- Rolling pin
- Double boiler

TIP If you don't have a purpose-made mould, you can use any household plastic container of a suitable size. You won't be able to thread the wick through the centre, so stand the thin candle the other way up, and cast the ice candle around it.

2 ▼ PREPARE THE MOULD Place the candle in the mould, threading its wick through the hole in the end. Secure the wick on the outside with mould seal, Blu-Tack or plasticine, keeping the candle central.

3 ▶ MELT THE STEARIN Use a wax double boiler or clean tin can in a shallow pan of boiling water to melt the materials without overheating them. Start by weighing out and melting 15gm of stearin, which is one-tenth of the quantity of the wax you will be using. Then chop a small amount of blue dye disc and add it to the melting stearin. You will only need about one-sixteenth of a disc, and remember that it is easier to add colour, rather than take it away, so don't overdo it.

4 ◀ **ADD THE WAX** Add the weighed wax granules to the dye mixture. Stir the mixture until the wax is completely dissolved and the colour is uniform.

> **TIP** Use an old metal spoon to stir the wax. Once the wax has melted, the spoon can be used as a good indicator of temperature, as the wax will coat the metal surface when it gets too cold.

5 ▶ **MELT THE WAX** Heat the wax to the correct temperature for pouring, checking with a thermometer. If you are going to use a pouring jug or metal ladle to transfer the wax to the mould, warm it by standing in hot water, otherwise it will cool the wax on contact.

6 ▲ **CRUSH THE ICE** Pack the ice cubes into a sturdy plastic bag – a supermarket carrier bag or a freezer bag will do. Secure the top and hit with a rolling pin to break up the ice. There is no need to make the pieces a regular size. Small pieces of crushed ice will produce a wax filigree effect. Larger chunks will produce larger holes in the candle.

7 ▶ **FILL THE MOULD** Working quickly before the ice has time to melt, pack the crushed ice into the mould around the core candle. Don't bring the ice any higher than about 1cm below the top edge of the mould (which will become the base of the candle). This will allow the wax to fill this area of the mould completely, creating an area of solid wax at the base of the candle, giving it stability. Ensure that the candle is standing in the centre of the mould once the ice is around it.

8 ▼ **POUR THE WAX** Work quickly to avoid the ice melting before you pour in the melted wax. Stand the mould on a dish to catch any drips and pour the hot wax into the mould immediately. Fill the mould right to the top, but do not be tempted to pour in more wax at this stage if the level sinks a little – allow the wax to harden in the gaps left by the melting ice.

Options...

Alternative materials

Although you will get excellent results from the step-by-step method shown here, you don't have to use the wax pellets, stearin powder and dye discs to make up the light blue wax that you need. Instead, you can simply melt down an existing candle. If you can't find a candle that's the right shade of light blue for your ice candle, melt some plain white candles together with a darker blue one to achieve the desired colour. If you don't have a full set of candlemaking equipment, you can also adapt the method given for coloured-layered cones which uses household items.

9 ▼ **EMPTY THE MOULD** Leave the wax to set hard – this will take two or three hours, during which you may need to top up the base with a little wax. Then hold the candle over a bowl and remove the mould seal so that the water runs out. Carefully remove the candle. Trim the wick at the base flush with the wax and trim the top wick to 1cm. If any of the core candle protrudes, this will soon burn down when lit.

floating candles

Here's how to make your own flower-shaped floating candles. To make four candles, you need coloured wax, stearin, wicks and a special double mould.

These delicate, little yellow floating candles are very easy to cast. The casting wax comes in the form of fine ready-coloured pellets, so that you don't need to add any dye, just the stearin powder which helps to give the candles the correct consistency and slow-burning characteristics. The finished candles can be used either singly or in groups, displayed in a bowl of water. They will naturally float just free of the surface as they burn down. For an even more striking display (as shown to the left), float a few flower petals on the water. It also looks very effective if you tint the water with a little food colouring or coloured ink.

Wax and wicks
The wax is provided in the form of pellets, making it easy to measure and melt. You also need some stearin powder, plus ready-primed wick.

1 ▼ **MELT THE WAX** Use a candlemaker's double boiler or a clean tin can in a pan of gently boiling water (see page thirteen for example) to ensure that you do not overheat the wax. To make two candles, pour half the wax pellets into the pan and allow them to melt.

Candle mould
The double mould you need is made from clear plastic, making it easy to see what's going on inside.

You will need:
- Yellow wax pellets
- Stearin powder
- Wick
- Double mould
- Candlemaker's double boiler,
  Or: a cleaned tin can and an inexpensive shallow pan
- Pouring jug

2 ▶ **ADD THE STEARIN** The packet of stearin powder contains the correct quantity to add to the full amount of wax. If you are only using part of the wax packet, measure out a similar proportion of the stearin powder and add it to the pan. Allow the ingredients to melt and mix thoroughly, taking care not to allow the pan to boil dry and overheat the wax.

3 ▼ **POUR THE CANDLES** Transfer some molten wax to a small jug or similar utensil in order to control the pouring process. If you use a ceramic or glass jug, warm it first to avoid over-cooling the wax. Carefully pour the wax into each hollow in the mould, until it is almost flush with the top.

4 ◀ **ADD THE WICKS** As the wax cools, it will shrink, forming a hollow in the centre of the mould. Cut two 4cm lengths of wick and insert a wick into each. If you do this while the wax is still slightly soft, you will be able to bed the end of the wicks in the base (using a thin stick to press them into place) to help the wicks stay upright.

5 ◀ **TOP-UP THE MOULD** Add more molten wax to the well in the centre of each candle, to fill it up flush with the top. Make sure that the wick stays upright in the centre of the candle by pushing it back into place if necessary.

No problem!

If you overfill the mould, don't worry, and don't try to clean up the surplus. Simply allow the wax to cool, then remove the candle from the mould in the usual way. Break off the surplus pieces of wax and trim round the candle with a craft knife.

Options...

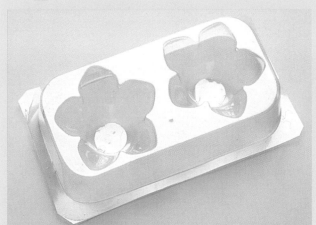

Making Extra Candles

The mould is long-lasting and can be used to make many more candles, simply by using extra materials. Just allow 25g of wax pellets and 5g of stearin per candle, plus a little wax dye in the colour of your choice. Use a 4cm length of primed wick for each candle.

TIP Don't try to hurry the cooling process. If you overcool the wax (for instance by putting the mould in cold water), it is likely to crack.

6 ▲ **REMOVE THE CANDLES** Leave the mould on a flat surface until the candles have completely cooled. As the mould has a wide taper, each candle should release easily, allowing you to lift it free by the wick. Under no circumstances should you pull hard on the wick, as this may cause it to pull out of the wax. If a candle is reluctant to move, warm the mould slightly in hot water to soften the wax.

decoratin
ca

g
candles

gilded leaves

It only takes a few minutes to decorate a plain pillar candle with these stylish gold panels. The 'gilded' effect is created with metallic gold wax, into which you can carve a unique design.

Sometimes, a simple, quick decoration can look considerably more stylish and sophisticated than a complex one. These small, simple, restrained gold panels are an example of an effective design that can be achieved in a mere few minutes.

The key to getting the subtle 'gilded' effect is to use a special metallic wax gilding paste, which is designed to leave a very thin, but very dense layer of colour on the surface. The results look much more realistic than gold paint, and although a pot or tube is quite expensive, it will last for ages. Two commonly available products from many craft and candle making supply shops include Treasure and Goldfinger.

You will need:
- Piece of white paper, roughly 8cm x 12cm
- Scissors
- Masking tape
- Pot or tube of metallic gold wax gilding paste
- Pillar candle 12cm tall
- Craft knife

Metallic wax
Sold in craft shops, art shops and some decorating suppliers, this is designed for gilded effects on all kinds of surfaces.

1 ▼ FOLD THE PAPER Cut a rectangle of white paper to the height of your candle (12cm in this case) and 8cm wide. Fold the paper into three along the 8cm length. Then fold it in half along the 12cm length. Cut a piece out of the middle of this fold, cutting 1.2cm in and 2.5cm along, parallel to the fold.

Masking tape
If you don't have any masking tape, ordinary adhesive tape can be used instead.

2 ▶ **MASKING THE CANDLE** When you open out the paper it should have three evenly spaced rectangular 'windows' as shown. Attach the paper to the candle using a strip of masking tape along each side.

3 ◀ **APPLY THE GILDING** Use your fingertip to rub metallic gold wax into the exposed squares of your template, working up to and over the edges of the cut-out squares.

4 ◀ CARVE THE PANELS
Remove the template from the candle by peeling away the masking tape. Then use a craft knife to make a leaf pattern across each gold square. First scratch the central rib diagonally across the square, then work the curves either side to form the leaf.

> **TIP** The beauty of the leaves is that they should look hand-drawn, not like perfect replicas of each other. However, if you are at all nervous about freehand drawing, simply draw a few examples first on a scrap of paper until you are comfortable with the shape.

5 ▼ FINISH THE LEAVES Make two little lines each side of the central rib to form the side veins.

Options...

'Negative stencilling'

Instead of cutting a mask that leaves small panels exposed, try this 'negative stencilling' effect. Simply cut small rectangular pieces of masking tape and stick them to the candle. Rub gold wax over the entire candle, then remove the tape to reveal the plain wax underneath.

warm hearts

Dress up any large candle in a matter of minutes with these bold heart motifs made from stick-on wax sheets. The finished results can be used just for decoration, but these heart candles will also make perfect Valentine's gifts.

Stick-on appliqué wax works very well with simple, bold motifs such as the heart shapes used here. They can easily be precut and placed where you want them, so long as you handle the soft wax with care.

It is easy to adapt the motifs to suit different sizes of candle. Both the examples shown here are decorated with 12 hearts cut from red wax and arranged in three rows of four. The smaller 15cm x 5cm candle is decorated with open hearts that have had their centres cut out, while the larger 15cm x 7.5cm candle has solid hearts arranged in a staggered pattern.

You will need:
- Appliqué wax sheet
- Pillar candle
- Soft pencil
- Craft knife
- Scissors

Appliqué wax
This specially formulated soft, sticky wax is supplied on a non-stick paper backing.

Pillar candle
Use a pillar candle at least 5 cm in diameter to suit the size of the motifs. Four hearts will fit around a 5cm candle with 7mm gaps between them, or around a 7.5cm candle with 27mm gaps.

1 ▶ TRACE THE HEARTS Use tracing paper and a soft pencil to trace the heart motif off the template on this page. Cut the shape out of a piece of thick paper to use as a pattern. Place the paper heart on to the backing paper of the appliqué wax and gently trace around it to transfer the pattern.

2 ▶ MAKE OPEN HEARTS To make the open heart shapes, carefully cut out the centre of the heart using a craft knife. To guide the cut, draw an inner line 6mm inside the line you have already marked. The pieces you cut out won't be wasted, as they are perfect heart motifs that can be used to decorate a smaller candle.

HEART TEMPLATE

3 ▶ CUT HEART SHAPES To cut out the main heart shapes, it is quicker and easier to use a pair of scissors. Hold the wax sheet gently as it is very soft and will mark easily.

4 ▲ APPLY WAX MOTIFS Peel off the backing paper and carefully position the hearts on the candle. If you lay them in place gently, you will be able to adjust their position a little. Press down firmly to bond them permanently in place.

appliqué wax motifs

Here's how to transform a plain candle into something special! The thin, coloured appliqué wax is easy to apply, and there is a choice of precut motifs or a sheet to cut to your own design.

This is the method to use for the gold stars and crescent moons, which come in the form of pre-cut wax motifs to provide instant, easy decoration for almost any type of plain candle. Feel free to experiment with the design – there is no need to copy the layout shown here. The only point to watch is to try and arrange the motifs neatly, because if they are applied too randomly, the result may look untidy and unfinished.

You will need:
- **Precut appliqué wax sheet (stars and moons pattern)**
- **Candle**
- **Craft knife**

Appliqué wax
The specially formulated soft, sticky wax is supplied on a non-stick paper backing.

1 ▶ REMOVE THE EXCESS
Lay the appliqué wax sheet down on your worktop, paper side down. Carefully peel the wax sheet away from one corner, leaving the motifs in place on the backing. This method ensures that you don't damage the soft motifs by trying to free them from the surrounding wax sheet individually.

Pillar candle
Other than a pillar candle as used here, any sort of candle is suitable, including coloured types. But try to avoid one that is too slim or oddly shaped, as this will distort the motifs.

2 ▶ SEPARATE THE MOTIFS
Lift off the individual motifs as you need them. You can keep those you don't need now (and the surplus sheet) for use in decorating another candle. It's a good idea to lift small motifs away using the point of a craft knife.

3 ▶ START THE DESIGN
Build up the design slowly. The wax will stay in place on the candle but won't bond firmly until it is smoothed down, allowing for repositioning. Rub over each motif with a fingertip when you are happy with its position.

4 ▲ FINISH THE DESIGN
The finished candle, ready for use. Because the appliqué motifs are themselves made from wax, they will melt and burn down naturally as the heat of the flame reaches them.

Cut out your own appliqué wax candle decorations from the plain sheet of appliqué wax, using the following method. You can either copy the Japanese symbols for which templates are given on the next page, or make your own unique designs. The key to getting neat results is to cut out the motifs on the candle itself, and then peel away the excess. In this way, there's no risk of thin pieces of wax breaking up when you try to transfer them. It's also much easier to keep all the elements of the design aligned with each other.

As with the gold stars and moons motifs, you can apply the decorations to the pillar candle provided, or to any other type of ready-made candle that appeals to you.

You will need:
- Plain appliqué wax
- Candle
- Tracing paper (or greaseproof paper)
- Soft pencil
- Craft knife

Pillar candle Similar choices apply as when choosing a candle to use with the gold motifs.

Appliqué wax sheet Sheets of paper-backed wax are available in a wide range of colours.

1 ◀ TRACE THE TEMPLATE Start by tracing your chosen design on to a sheet of tracing paper. Scribble over the back with a soft pencil, then tape the trace over the backing-paper side of the wax sheet. Now go over the traced lines to transfer them to the backing paper. Cut out a rectangle of wax sheet, leaving a small border around the motif.

2 ▼ CUT OUT THE DESIGN Lightly press the wax sheet into position on the candle. Then go over the lines with a craft knife, making sure you don't miss any, so that the knife cuts right through both the paper and the wax itself.

3 ▲ REMOVE THE EXCESS Carefully peel away the wax sheet, starting from one corner. If the motif starts to lift, press it lightly back into place with your finger. If it won't come free, you may have to cut the wax sheet a little more.

TEMPLATE COLLECTION – **JAPANESE SYMBOLS**

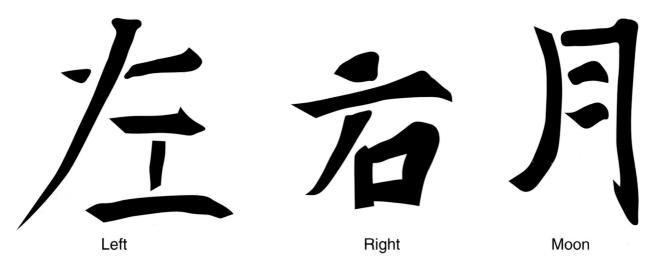

Left　　　　　Right　　　　　Moon

Harmony

4 ▼ **PEEL OFF THE BACKING** Rub over the motif with the end of your finger to bond the wax to the candle. Then use the point of your craft knife to peel away the paper, taking care not to mark the surface of the wax or the candle.

carving dipped candles

Many coloured candles are made by dipping a plain candle to give it a thin coating of coloured wax. This makes it easy to form decorations by cutting away the layer of colour to expose the underlying layer.

Carving a dipped candle with very simple geometric shapes provides an easy way to produce an attractive pattern in a few minutes. You must use an over-dipped candle in order to reveal the contrasting colour of the wax beneath. When you burn the candle, the light will glow through the carved pattern as it burns down, so choose a simple wooden or glass candlestick for best effect.

You will need:
- Colour-dipped tapered candles
- Craft knife

Colour-coated candles
It should be easy to detect a coated candle, rather than a solid-colour one, by looking at the base.

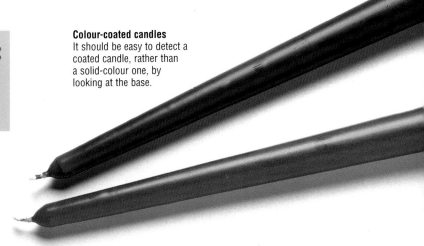

1 ▼ SCRATCHING THE SURFACE Starting about 3cm above the base, use a craft knife to draw a simple geometric shape, such as a triangle, on to the surface of the candle. Do not carve too deeply: the dipped layer will only be about 2mm thick. Don't worry if the wax breaks up within the shape, as you will be removing it.

Options...
Alternative Patterns

You can vary the design to suit yourself, but don't try to be too ambitious when making freehand cuts with a craft knife. Even if you can make the cuts accurately, it will prove difficult to remove the surplus wax from within a fiddly pattern. Bolder teardrop shapes, wiggly lines and scrolls all work well. Choosing the right colours can be just as effective as the design of the carving. For example, make a group of red and dark-green candles for the festive season, with a stylised holly leaf design a few centimetres above the base of the candles.

2 ▶ EXPOSING THE BASE CANDLE Use the craft knife to lift the wax out within the cut outline. You may have to pick out the pieces of coloured wax in stages, recutting as necessary.

3 ▶ CARVING THE CANDLE Continue cutting a series of geometric shapes, working all around the candle, upwards from the base. Make the shapes around the top a little smaller than those lower down.

glitter tealights

Cheap-and-cheerful tealights can be used for hundreds of candle projects. But what could be simpler than turning the tealights' own metal case into a decorative holder?

Tealights consist of a candle and holder all in one. But although the metal case does not look particularly attractive, it only takes a few minutes to decorate it with glitter and make these fun, sparkly holders. The only thing to watch out for is that the glue you use must be able to withstand the heat of the burning candle without melting or giving off unpleasant fumes. For this reason, we used a hot-melt glue gun, although other types of glue can be used instead.

Make sure that you stand the tealights on a heatproof surface before lighting them, and take care not to touch the cases immediately after extinguishing the flame, as they can become quite hot.

You will need:
- Tealights
- Hot-melt glue gun and glue sticks
- Selection of glitter, sequins, beads and other sparkly shapes

Glitter and sequins
Buy a selection of different colours and shapes to add variety to your tealights.

1 ▶ PREPARE THE CASES
It's better to remove each tealight from its metal case, as there's no risk of getting glitter on the wax itself. Taking care not to squash the thin metal, apply glue to a small section of the case. If the candle won't come out of its container, simply hold the tealight by its wick and take care not to get glue or glitter on the wax as this will look untidy.

2 ▶ APPLY GLITTER
Before the glue has time to dry, sprinkle glitter on to a sheet of clean paper and roll the glued section of the case in it. Apply glue to the next section of the case and repeat until the whole case is covered with glitter.

SAFETY!
If you use a glue gun for this project, you will find that the thin metal of a tealight case conducts the heat of the glue very quickly and may become hot, so handle it with care, wearing gloves if necessary.

Options...
Ringing the changes
It's easy to create different patterns by using alternative types of glitter shapes. For example, after you have rolled the tealight case in the first layer of glitter, you can apply glue around the top of the case only and roll it again, this time in small glitter stars.

3 ▶ ADD STARS
Put a dab of glue halfway up the side of the case and stick on a sequin flower. Repeat at intervals around the case. Allow the glue to set completely before returning the candle to the metal container.

47

holders, la
co

Lanterns & Containers

shine a light

Take recycling to a new level and turn empty drinks cans into desirable objects by using them to create candle sconces. The shiny surface will act as a perfect natural reflector for the flickering candlelight.

Stripping the paint off an ordinary drinks can reveals a beautiful shiny surface into which you can cut out a scalloped design that will reflect the candlelight. If you place a tealight in a deep sconce, it will give a subtle glow, while taller candles will give a stronger light.

The thin metal of a drinks can is easy to cut with an old pair of scissors, and you can leave the pointed opening plain or punch in a pattern using a nail, hammer and chisel. The finished sconce can be placed on a shelf or hung from a wall, and is suitable for inside or outside use. As the metal can get quite hot, do not stand the sconce on a surface that could be damaged by heat.

You will need:
- **500ml can. Cans may be aluminium or steel (tinplate). Soft, flimsy aluminium does not work as well as steel.**
- **Gel paint stripper**
- **Kitchen paper**
- **Rubber gloves**
- **Old pair of scissors**
- **Pencil**
- **Tracing paper**
- **A4 sheet of plain paper**
- **Masking tape**
- **Fine marker pen**
- **Nail**
- **Hammer**
- **Chisel**

1 ▶ STRIP THE CAN Pour a little paint stripper on to a pad of kitchen paper or cloth and gently rub the surface of the can – you will find that the paint rubs off easily. Protect your hand with a rubber glove, as stripper may irritate your skin.

2 ◀ REMOVE THE CAN TOP Use an old pair of scissors to cut the top off the can. Stab one blade into the metal and roughly cut around the top just below the curved section, taking care with the sharp edges.

3 ▶ MAKE THE TEMPLATE The diagram (right) is for half the template. Trace it off the page using a piece of thin paper or tracing paper. Place the trace on to a folded piece of paper, with the indicated edge along the fold. Cut along the opposite edge and unfold the paper to give the complete template, which will fit a standard drinks can. Wrap it around the can, overlapping the ends and keeping the upper edges level.

SAFETY!

Watch out for the cut edges of the can, especially when bending the points, which can be sharp. If you are worried about pricking a finger, wear heavy work gloves.

TIP The same template can be used to make a shorter sconce to suit a tealight. Either move the template lower down the can and cut off the surplus metal left at the top, or use a shorter, 330ml can.

PLACE THIS EDGE ALONG FOLD IN PAPER

METAL LANTERN TEMPLATE TO SUIT 500ml CAN

4 ▶ MARK THE OUTLINE Use a fine marker pen to trace around the scalloped edges of the template on to the can. Test the marker first – you don't want the line to rub off while you cut the can. Remove the template from the can and cut away the bulk of the metal, a little way above the marked line, using an old pair of scissors.

5 ▼ THE FINISHING CUT Work around the can again, this time cutting carefully around the lines of the design. It's easiest to start at the point of each triangular shape and snip down each side in turn, rather than trying to cut in and out in one go.

6 ▶ **BEND THE POINTS** Leaving the largest point at the back upright, work around the can and bend back the triangles at an angle to the can using your fingertips. Take care when handling the sharp, cut edges of the metal.

7 ◀ **APPLY A PUNCHED DECORATION** Rest the points on a pad of soft cloth or kitchen roll over a chopping board or scrap of wood, and use a hammer and nail to punch a single hole near the tips of the back point and the two points on each side.

8 ▼ **CHISELLED FEATURES** For extra decoration, use a chisel to make three straight cuts in a fan shape in each of the five main points. Work over a pad and board as before and tap the handle of the chisel sharply with a hammer.

Options...
A pierced front pattern

You can continue the pierced pattern on the front of the can, so long as you support the metal internally to stop it buckling on impact. The easiest way to do this is to fill the base with water and leave it in the freezer. When the water has frozen, you can use the nail to hammer a pattern of dots into the front of the can. Work quickly, before the ice melts.

orange pomander holders

These simple shelf or table decorations are based on the fragrant pomander balls, made from oranges and cloves, which are traditionally used to scent cupboards and drawers.

These holders are made by scooping out a small hole in the top of a fresh orange and simply popping in a tealight. Surround the hole with cloves, which will emit a subtle fragrance, especially when they are warmed by the candle.

As the main picture shows, it's easy to make different decorative effects by adding extra rows or scallops of cloves. The pomander should last for a few days, and can be reused by changing the tealight. You can even recycle the cloves to make a new pomander.

You will need:
- Several unblemished oranges
- Ballpoint pen
- Tealights
- Cloves
- Sharp knife

Oranges
Use fresh oranges with a firm skin and a thick peel.

1 ▼ MARK THE ORANGE Make sure that the orange will stand firmly on its base, then place the tealight on top of the orange and draw around it with a ballpoint pen.

2 ▼ CUT THE HOLE Using a sharp knife, cut into the peel along the line. Scoop out enough flesh to accommodate the tealight, making a hole around 2cm deep. The tealight should be quite a tight fit.

Cloves
Buy an ordinary pack of dried, whole cloves from the spice counter.

3 ▼ ADDING THE CLOVES Push the tealight into the orange – you may need to use both thumbs – and stud around with cloves. Then add decorative lines to the rest of the orange.

Options...

Ring the changes

Oranges are not the only fruit that can be used as candleholders. The method used to make the pomanders will work just as well with apples, although the scent will be subtly different because oranges tend to add citrus overtones to the smell of the cloves. Use a very firm, hard apple such as a Granny Smith. Your holders will last even longer if the apples are slightly unripe.

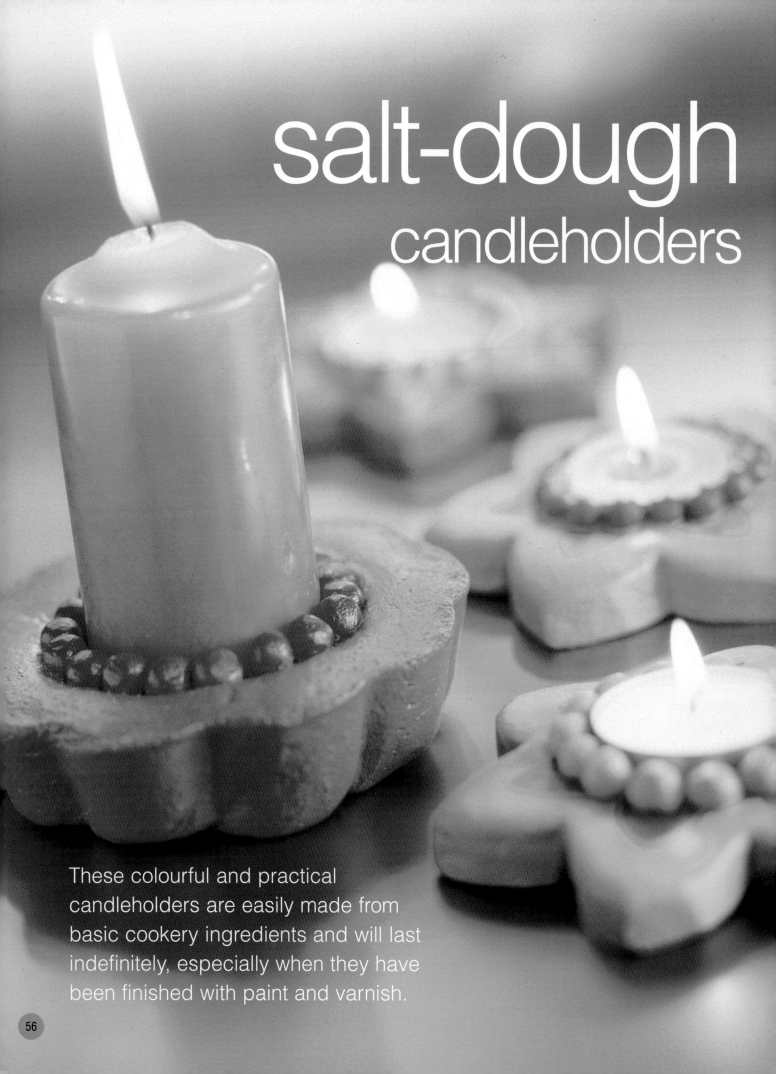

salt-dough candleholders

These colourful and practical candleholders are easily made from basic cookery ingredients and will last indefinitely, especially when they have been finished with paint and varnish.

Salt dough makes amazingly hard and durable objects – these holders are very strong due to the lengthy baking time in the oven. Use a small, shaped cake tin as your mould for a candleholder to take a larger candle, or use the template supplied to make smaller holders for tealights. The basic method is the same and you can have fun decorating the finished holder with acrylic paints in sizzling colours. The example made here is ideal for a candle that is 5cm in diameter and 12cm or so tall, but you can easily adapt the method to suit a different-sized example.

Paint
Small pots or tubes of artist's acrylic paints are an ideal way to produce the vibrant colours shown. You can also use tester pots of emulsion paint, which are cheaper, but you are unlikely to find such strong colours.

1 ▶ MIX INGREDIENTS Put the plain flour and salt into a bowl and add half the water. Use a spoon to stir the mixture and blend the ingredients.

Cake tins
Small, shaped tins sold in hardware stores and kitchen shops offer a variety of patterns suitable for making candleholders. Choose non-stick coated ones if you have the option.

2 ▶ MAKE THE DOUGH Gradually add the remaining half cup of water and knead the mixture with your fingers for about 5 minutes until you have a pliable dough – turn it out onto a wooden board if you find this helpful. Set it aside in an airtight container for 10 minutes to develop pliability.

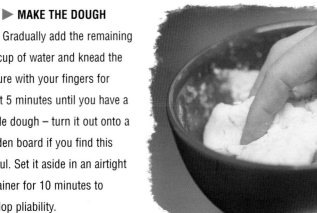

You will need:
- 2 teacups plain flour*
- 1 teacup salt*
- 1 teacup water*
- Acrylic gesso or white emulsion paint
- Artist's acrylic paints
- Clear satin polyurethane varnish
- Spoon
- Airtight container
- Flower-shaped cake tin 13cm diameter x 3.5cm deep
- Fine sandpaper

*The quantities given will make four candleholders.

3 ▼ PREPARE THE MOULD Smear the inside of a small flower-shaped cake tin with lard or cooking fat to prevent the dough sticking to the surface.

4 ▶ FILL THE MOULD Divide the dough into four and push one generous portion into the tin. Press it down firmly with the flat of your hand to ensure that it fills out the corners of the tin.

5 ▶▼ **LEVEL THE DOUGH** Hold a sharp knife horizontally and cut across the top of the dough to level it. Then moisten your fingertips and smooth over the surface of the dough.

6 ▼ **MARK THE HOLE** Stand your chosen candle in the centre of your salt dough and trace around it with the tip of a knife. Remove the candle.

7 ▶ **CUT THE HOLE** Using a knife held vertically, cut straight down through the dough, working slightly outside the marked line. When you have gone all the way round a couple of times, remove the dough 'plug' from the centre with the point of the knife. You will find it comes away quite cleanly.

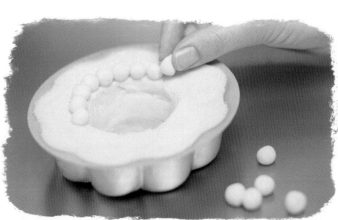

8 ▲ **TRIM THE HOLE** Smooth the cut edges of the hole with a moistened fingertip. Then roll some of the dough into small balls measuring about 1cm across. Moisten around the edges of the cut-out and press the balls in place, next to each other. If necessary, use the candle to check that the balls won't obstruct the hole, but don't worry about any uneven edges around the top of the tin as these can be sanded away later.

9 ▼ **BAKE THE DOUGH** Bake the candleholder for approximately 7 hours at 120 deg C/250 deg F/Gas Mark 1/2. Halfway through the baking time, remove the mould from the holder. After the full baking time, turn off the heat and allow the candleholder to cool in the oven.

10 ▼ **SMOOTH THE SURFACE** Gently sand all surfaces of the candleholder with fine sandpaper. If you accidentally detach any of the balls, they can be glued back in place with epoxy resin adhesive.

11 ▶ PAINT WHITE

Undercoat the candleholder with acrylic gesso or white emulsion paint and leave to dry for the amount of time specified on the pot.

12 ▶ PAINT IN COLOUR

When dry, paint with acrylic paints – we used shades of red and shocking pink. When dry, apply several coats of clear polyurethane varnish, allowing the holder to dry fully in between each one.

tealight holders

You can make a variation of the candleholder to suit a tealight. This time, you don't need to use a mould, as you simply roll a thick slab of dough and cut out the shape of the holder using the template supplied.

1 ▶ PREPARE THE DOUGH To make four holders, make up half the quantity of dough used for the main project and knead it well. Divide it into four and roll one portion out flat, 12mm thick, on a sheet of baking parchment.

2 ▼ CUT OUT THE SHAPE Trace off the template and cut out the shape from a piece of stiff card. Place the card template on top of the dough, hold it down with your fingertips and cut around it using a sharp knife held vertically.

3 ▼ FINISH THE HOLDER Peel away the surplus dough from around the cut-out. Set a tealight in the centre, mark out and cut out the central hole as before. Moisten your fingertips to smooth out the dough around the edges and in the centre, then bake in the oven (120 deg C/250 deg F/Gas Mark 1/2) for 6 hours.

4 ▶ PAINT THE HOLDER Follow steps 11-12 above, painting the candleholders in bright shades of turquoise, orange, yellow and pink.

mosaic votive candle holder

Votive candles were originally used in religious ceremonies. This decorative holder uses colourful mosaic tiles, glass beads and gold paint to simulate the sumptuous feel of precious, jewelled originals. When a candle is lit inside, its light will glow through the surface.

This holder is easily made by gluing bands of mosaic tiles in co-ordinating colours, enhanced by jewellery stones, to a glass. Choose a frosted glass if possible, as it will look neater on the inside if the adhesive and tiles can't be seen through the glass. You can leave the grout lines plain white or coat them with gold acrylic paint for a particularly sumptuous look.

The method shown is based on leaving the first row of tiles and beads to harden before finishing the design. Although this means waiting after Step 3, it has the advantage of giving you a firm base for holding the glass while you build up the design. If you prefer, it is possible to apply all the tiles in one go, but you will have to take care not to dislodge them as you work.

Tile adhesive & grout
This type of product is designed both to stick the tiles in place and fill the gaps. Buy a small tub from a DIY store.

Frosted Glass
You can use either a small drinking glass or a glass that is designed as a candleholder.

Tesserae mosaic squares
These colourful 2cm glass squares are available in a wide range of shades from DIY stores and craft shops.

Gold acrylic paint
You only need very little, so buy the smallest pot from an art shop or DIY store.

Jewellery stones
A wide selection is sold by specialist bead shops, or can be bought by mail order.

You will need:
- Frosted glass, 7cm tall
- Ready-mixed combined wall tile adhesive and grout
- Spreader
- Tesserae mosaic squares, 2cm square, in three co-ordinated colours
- One cabouchon jewellery stone
- Smaller jewellery stones
- Kitchen paper
- Fine paintbrush
- Gold acrylic paint

TIP Before starting to stick your mosaic tiles in place, try laying them around the glass to get an idea of how they fit and how wide the gaps between each row should be.

1 ▶ APPLY THE ADHESIVE
Use a spreader to apply a 3mm layer of tile adhesive to the outside of the glass in a vertical strip about 5cm wide from the top to the base.

2 ▼ **START THE DESIGN** Arrange two tesserae squares in a vertical row on the adhesive, with a cabouchon jewellery stone in the centre. Place two small jewellery stones near the lip of glass to form the start of the row. Press the stones and mosaic into the adhesive.

3 ▼ **REMOVE EXCESS ADHESIVE** Use the spreader to remove excess adhesive around the edges of the row before it hardens. Any smears on the tiles can be wiped off with a pad of damp kitchen towel.

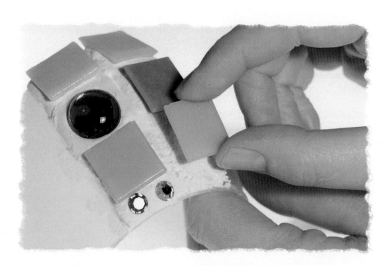

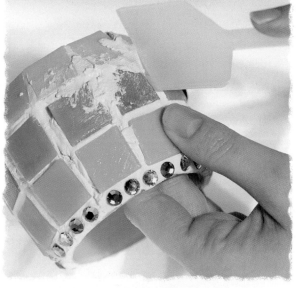

4 ▲ **ADD A SECOND ROW** It's a good idea to leave the first row to harden for about an hour before applying more adhesive and building up the second row of the design, this time using a mosaic tile in place of the cabouchon stone. In this way, the first row will form a firm base for the pattern and avoid any problems caused by trying to apply too many pieces of mosaic in one go before the adhesive hardens – such as dislodging the tiles you have already positioned.

5 ▲ **FILLING THE GAPS** When all the mosaic tiles and jewellery stones are firmly stuck in place, use the spreader to apply a little more adhesive over the surface, working it into the spaces between them.

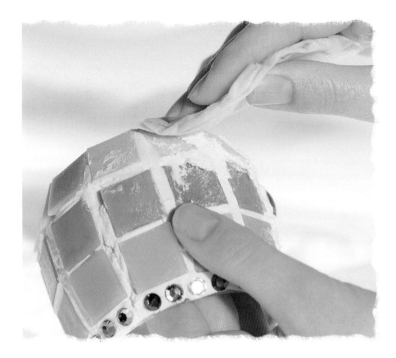

6 ◀ **POLISH THE TILES** Fold a double layer of kitchen paper to form a pad and dampen it with water. Wipe the pad over the surface of the tiles and jewellery stones to remove the excess grout and restore the shine to the surface.

7 ▶ **PAINT THE GROUT** Leave the grout to harden overnight, then use a fine brush to paint the lines between the tiles gold using acrylic paint. Leave to dry.

Options...
Broken China mosaics
Instead of mosaic tesserae, you can adapt the idea by using broken pieces of coloured crockery to form an irregular pattern. However, these will not allow the light to shine through.

chinese paper lanterns

Here's a candleholder that looks really good unlit, but comes to life when the flame of a tealight burning inside illuminates the delicate shades of the paper covering.

Oriental papers have very distinctive styles and rich colours that complement candlelight perfectly. With a wide range to choose from, you can make as many individual holders as you like, either for everyday decoration or special occasions.

Beautiful papers don't even need to cost very much. Some of the examples shown on the right are actually taken from food wrappers. The only items you need are a suitable drinks glass and a raffia tie to keep the paper in place around it.

You will need:
- Tealight
- Oriental patterned paper
- Raffia ribbon
- Glass tumbler
- Adhesive tape

Oriental patterned paper Suitable papers can be found in oriental food stores as well as some craft shops

Raffia ribbon Choose a bright colour. Natural raffia will stand the heat of a candle better than plastic gift ties.

Drinks glass A short, wide tumbler is the ideal choice for this project.

1 ▶ MEASURE YOUR GLASS
Use a tape measure to check the height of your glass, then measure around the base. Add 1cm to the second measurement to ensure that the paper will overlap when you wrap the glass.

2 ▼ CUT THE PAPER
Cut a rectangle to these dimensions, placing it to suit the pattern on the paper (see Options).

3 ▼ COVER THE GLASS
Wrap the paper around the glass, securing the overlap with a small piece of adhesive tape in the centre, where it will be hidden by the raffia tie.

4 ▶ FINISHING TOUCH
Use several strands of brightly coloured raffia to tie a large bow around the middle of the glass, holding the paper in place. Trim the loose ends of the raffia neatly after pulling the bow tight.

Options...
Different Patterns

The enormous variety of papers available means that you are spoilt for choice, and every holder will be individual and different. The only thing to consider is that when you position the measured rectangle on the paper, make sure that it aligns with the pattern. By positioning it carefully, you can achieve effects such as the panelled motif seen far right, or a geometric border as seen to the left.

creating displa

radiant blooms

The key materials required for this simple yet decorative table centrepiece are a box of standard household candles and a bunch of seasonal flowers. The display only takes a few minutes to make, and as all the materials are easily obtainable, it can be put together at short notice.

This table decoration is easy to create without using any hard-to-find materials. It is based on a block of floristry foam (such as Oasis), held in a colourful bowl that tones with the flowers chosen. We used blue cornflowers in a 15cm pale yellow bowl with a deep, hemispherical shape, but you can easily adapt the foam filling to suit a slightly different size or shape of bowl. The candles will burn for up to 5 hours. Although the damp foam means that there is no fire risk, you should extinguish the flames before they burn down to the level of the flowers, in order to stop the petals charring.

Seasonal flowers
Choose flowers such as cornflowers or carnations that have small, compact heads and reasonably firm, thin stalks. There is no need to buy costly specialist blooms.

Floristry foam
A block measuring 23cm x 11cm will provide plenty of material for a deep bowl 15cm in diameter.

Masking Tape
This will be covered up in the final finished decoration.

Household candles
Using a box of standard white household candles means that they will work with any colour of flowers and you do not have to buy them specially.

You will need:
- Floristry foam (approx. 23cm x 11cm x 7cm)
- Kitchen knife
- Bunch of seasonal flowers
- One box of household candles
- Floristry wire
- Masking tape
- Wire cutters
- Kitchen scissors

1 ▶ PREPARE THE FOAM Slice the block of foam in half through its thickness, using a large kitchen knife. Place both halves together and turn the bowl upside down on top **(inset)**. Draw around the bowl to mark a cutting line.

2 ▶ **CUT THE FOAM** Use the knife to cut the two marked semicircles out of the blocks of foam, cutting at a slight angle so that they will fit into the top of the bowl.

3 ◀ **ADD FOAM TO THE BOWL** Cut a block from the surplus foam so that it just fits inside the bottom of the bowl. This is to support the two semicircles flush with the top of the bowl. (If your bowl is shallower than the one shown, you may need to trim the block until the foam is flush with the top.) When all the foam is in place, add water until it is thoroughly soaked.

4 ▶ **MAKE THE CANDLE SUPPORTS** Cut four 5cm lengths of floristry wire and tape them around the bottom of each candle, leaving about 3cm protruding, at equal intervals. Use two or three turns of tape to ensure that the wires are firmly fixed in place. Repeat until all six candles have four wires attached.

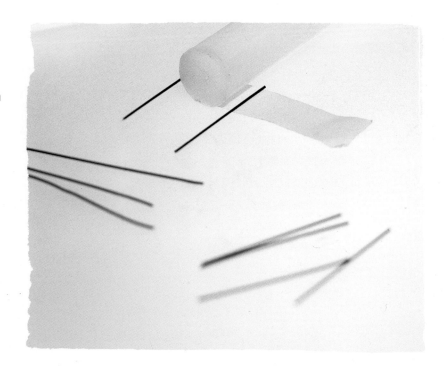

5 ▶ **ADD THE CENTRAL FLOWERS** Cut off the flower heads, leaving about 1.5cm of stalk showing. Press the stalks into the foam base near the centre, working around so that the flower heads are close enough to conceal the foam completely. Carry on building up the floral display until you have filled an area 10cm in diameter in the middle of the bowl.

> **TIP** If the flowers don't have very strong stalks, you may find it hard to push them into the foam. But if you use a cocktail stick or similar tool to make a hole first, you will find it much easier to insert each stalk.

6 ▼ **POSITION THE CANDLES** Place a candle at the edge of the ring of flowers, about 1cm from the edge of the bowl, and push it down so that it is held upright by the wires. Repeat for all six candles, making sure that you position them at a similar distance from the edge and evenly spaced around the bowl.

7 ▼ **ADD THE OUTER FLOWERS** Carry on building up the floral display around the candles until all the foam is covered, making sure that the tape around the candle is also concealed.

giftwrap candle cones

A great way to provide a present with a personal touch is to giftwrap a candle you have bought or made. Although it looks like a traditional sweet cone, this parcel holds a single candle as a delightful surprise!

Candles always make attractive presents – especially if you made them yourself. And to make the gift even more special, here's an unusual wrapping that will make its contents a complete surprise. As it is based on a conical shape, this giftwrap suits cylindrical candles, with or without points. The size shown here suits a medium-sized pillar candle (or a bunch of thin tapers) but is easily scaled up simply by enlarging the dimension that you set at Step 1. There's also plenty of scope for you to vary the materials and produce a host of different, individual wrappings.

Card or handmade paper
Choose a stiff paper that will roll easily without cracking.

Ties
Use a pretty ribbon, braid or parcel tie as you prefer.

Tissue paper
Buy two sheets of colour-coordinated giftwrap tissue.

Clear or iridescent film
Decorative film for giftwrapping is normally sold in rolls. You can use either clear or patterned film, so long as you can see the wrapping paper through it.

You will need:

- Piece of thin card or thick handmade paper, about 18cm square (depending on the size of your candle – our example is 10cm tall and 3.5cm wide)
- Pair of compasses
- Scissors
- Sheet of clear or iridescent film
- All-purpose adhesive
- Pencil
- 2 sheets of tissue paper
- Ribbon, braid or raffia for ties

1 ▶ DRAW A QUADRANT Using a pair of compasses set to a radius of 18cm, position the point on one corner of the paper or card and draw a quarter circle.

73

2 ▶ **CUT OUT THE CONE** Use a pair of scissors to cut the paper along the marked line.

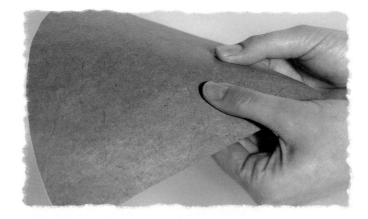

3 ▲ **FORM A CURVE** Start to roll the paper just above the point between thumb and finger to soften the paper and help curve it into a cone.

4 ◀ **GLUE THE CONE** Holding the cone shape together, apply a little glue in a strip along one of the straight edges, then overlap the other edge and press together. It helps to hold a pencil inside the cone behind the seam to act as a support. Work along the seam, until the full length is glued.

5 ▶ **CUT TISSUE STRIPS** Use shredded tissue to pack out the cone and protect the candle. To cut the strips, place the two sheets of tissue paper together and fold them into four layers. Then use your scissors to cut the wad of paper into narrow strips about 5mm wide.

6 ◀ **FILLING THE CONE** Pull the strips of tissue apart and shake them together to form a loosely tangled mass. Push some of the shredded tissue into the point of the cone, then put the candle inside and pack more shredded tissue around it.

7 ▶ **WRAP THE CONE** Place the cone on the corner of the piece of film and roll it so that the edges meet over the seam in the paper. Secure the film with a dab of glue or clear tape at the top of the seam.

8 ▼ **TIE THE TOP** Gather the film around the top of the cone and tie with a length of ribbon, cord or raffia. Cut off any surplus to form a neat frill.

Options...

Using fabric

Instead of wrapping the cone in film, you can use an attractive sheer fabric. Suitable choices include organza, available in a range of wonderful colours, or the type of netting used for veils and petticoats, which comes in a range of shades from subtle to wild! Tie the top with fabric ribbon if you prefer.

pastel glow

Stunning table displays don't have to be complicated or expensive. This arrangement of flower heads and tealights makes a perfect alternative to a vase of flowers and a set of candlesticks.

Combining flowers and candles makes an effective table centrepiece for a dinner party or an occasional display – but many arrangements depend on special vases or candle holders to prevent the candles scorching the flowers. However, this display requires no more than a shallow dish, plus a bunch of seasonal flowers chosen to give a burst of bright colours. The tealights used will burn for around four hours, while the flowers should last long enough for the candles to be replaced so that the display can be used for more than one day.

You will need:
- Around 20 flower heads, depending on their size, and the dish
- Shallow dish, around 20-25cm in diameter
- Tealights (nightlights)
- Kitchen scissors

1 ◀ PREPARE THE DISH Fill the base of the dish with a little water to a depth of around 1.5cm and place three tealights near the centre. Do not make the water any deeper, or the tealights may start to float.

Flowers
You can choose any sort of flowers so long as they have a densely packed head of petals between 3-5cm in diameter. This example uses ranunculus and globe flowers.

Tealights
Choose between low-cost white tealights or coloured ones for a more striking effect.

2 ▶ TRIM THE FLOWER HEADS Using kitchen scissors, clip the stems of the flowers just below the petals. Depending on the flowers you use, you may need to pick off any leaves around the flower heads.

3 ▼ POSITION THE FLOWERS Pack the flowers tightly around the tealights with their cut stems down in the water. Vary the colours of the petals and don't leave any visible spaces in the completed arrangement.

Options...
Using coloured tealights
For an even more striking effect, add coloured tealights, chosen to complement the colours of the flowers that you are using, so that they disappear into the arrangement.

black magic

This dramatic but simple arrangement is lit by the flickering flames of tightly packed tealights, concealed in the depths of a bed of coal-black pebbles and arranged on a stark metal plate.

With the candles all but concealed in the depths of a bed of dark stones, this dramatic arrangement has the same fascination as the flickering flames of an open fire, and makes a perfect decoration for a bare hearth or even for an unusual table arrangement. Almost any type of aluminium plate is suitable as a base, but avoid using stones from your garden as some types may shatter when heated. As with any candle display, don't try moving the plate when lit, and allow it to cool down after use, as it will remain quite hot for a considerable time.

Chicken wire
You only need about 20cm of fine-mesh chicken wire, available from a hardware store or garden centre.

Tealights
Buy a bag of low-cost tealights (or nightlights). The exact number you will use depends on the size of your plate.

1 ▼ CUTTING THE CHICKEN WIRE
Flatten out a piece of fine chicken wire and cut a rough circle to fit just inside the well in the centre of the plate, using wire cutters or an old pair of kitchen scissors.

2 ▼ POSITIONING THE TEALIGHTS Pack the well in the centre of the plate with tealights, leaving as few spaces as possible. This example uses 16 tealights, but the number you will be able to fit in depends on the diameter of the plate.

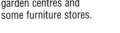

Black pebbles
Small bags of coloured pebbles are sold in garden centres and some furniture stores.

You will need:
- Metal dinner plate (see main text)
- Pebbles
- Tealights (the exact number depends on the size of the plate)
- Fine chicken wire
- Wire cutters

3 ▼ FITTING THE WIRE GRID Lay the chicken-wire disc on top of the tealights and raise the wicks through the gaps in the wire mesh using the end of a bamboo skewer or a similar tool. Then carefully lay the pebbles in place so that they are supported by the chicken wire. Make sure you leave gaps around each wick, although the stones can overlap the tealights a little.

4 ▶ COMPLETE THE DISPLAY
When all the stones are in place, the display is ready for use. Be sure you place it on a heat-proof surface, as the metal plate will become hot. Light the individual wicks using a taper, rather than matches.

metal-foil candle collars

Candle collars made from metal foil add a striking touch to plain candlesticks, for display or a special occasion. Emboss the foil with delicate scalloped patterns, or further embellish the designs by adding drop beads at the tip of each petal.

These candle collars are quite easy to make from the thin metal foil sold for craft work. Suitable foil is available from craft suppliers. However, it's also possible to make collars from recycled drinks cans. Use aluminium cans and remove the colour from the outside with paint stripper. Wash out the can, and use an old pair of scissors or some tin snips to cut off the top and base, stabbing into the can to start the cut. Take care not to cut your fingers while you do this. Cut vertically down the remainder of the can to produce a rectangle of metal, which you can reverse-roll to flatten it before use.

You will need:
- Tracing paper
- Pencil
- Thin card
- Fine (36-gauge) metal foil
- Old pair of scissors
- Kitchen paper
- Dried-up ballpoint pen
- Small cabouchon jewellery stones
- Clear all-purpose adhesive

Optional
- Small beads
- Headpins or fuse wire
- Jewellery or round-nose pliers
- Darning needle

Metal Foil Aluminium foil is sold in several gauges (thicknesses). You need 36 gauge.

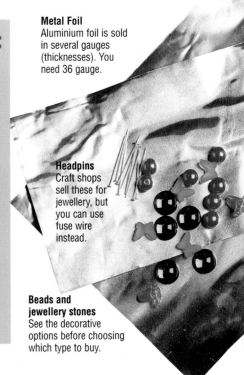

Headpins Craft shops sell these for jewellery, but you can use fuse wire instead.

Beads and jewellery stones See the decorative options before choosing which type to buy.

1 ▼ PREPARING THE TEMPLATE Trace off the template (below right) using tracing paper or thin paper, after testing the size against your candleholder (see Tip). If you use tracing paper, transfer the design to a piece of thin card and cut it out. Then draw around the template on to a piece of metal foil, using a felt-tip pen.

2 ▼ CUTTING THE SHAPE Using an old pair of scissors, cut out roughly around the drawn shape. Then cut along the outline, working from the points inwards on each petal shape.

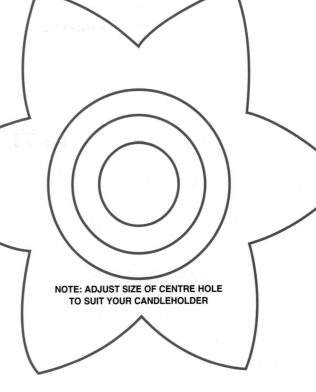

NOTE: ADJUST SIZE OF CENTRE HOLE TO SUIT YOUR CANDLEHOLDER

CANDLE COLLAR TEMPLATE

3 ▶ CUTTING THE HOLE Use the scissors to punch a hole through the centre of the candle collar, then cut out the central circle, working round it as smoothly as possible.

TIP The size of the socket into which the candle fits varies from one holder to another. The template shown is designed to suit common patterns of metal and glass holders and has several circles in the centre. These allow you to make the hole in the collar larger or smaller. To check which size you need, hold your candleholder over the template and note which circle is just visible around the socket.

Decorating your candle collars

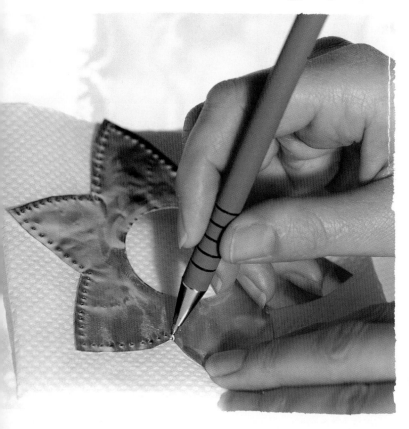

1 ◀ **ADDING AN EMBOSSED EDGE** Rest the collar on a folded pad of kitchen paper. Use a dried up ballpoint pen to make a pattern of dots all around the outside, close to the edge. Hold the pen upright and press down, spacing the dots as evenly as you can. Note that the side you are working on will become the underside of the finished collar.

2 ▼ **ADDING DECORATION** Here's how to add a pattern based on sticking jewellery stones to the 'petals' of the collar. Loosely position the stone where you want it, then draw in a simple design of regular squiggles using a ballpoint pen. Repeat for each petal.

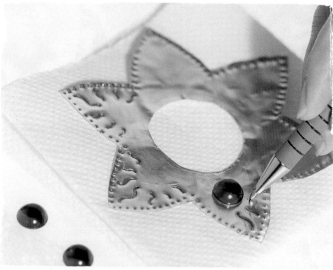

3 ▲ **APPLYING THE DECORATIONS** You have been working on what will become the underside of the finished collar. Now turn the collar over to the opposite side. Apply a small amount of glue to the back of each stone and set it in position on the collar.

Options...

Curled Petals

After embossing, roll the tip of each petal around the body of a pencil to curve them upwards. Take care only to press the metal gently, so that you don't flatten the embossed pattern.

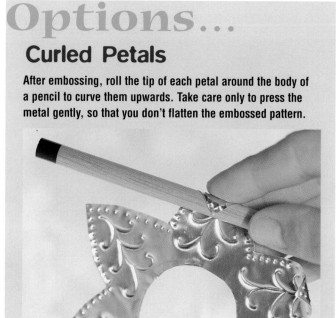

...and adding drop beads

1 ▶ CUT THE HEAD PIN Cut out a candle collar and emboss with a design of your choice. Thread a bead onto a headpin, then snip the headpin 12mm above the bead with an old pair of scissors.

2 ▼ CREATING A HOOK Bend the wire above the bead into a hook, using jewellery pliers or the tips of the scissors.

TIP As an alternative to the headpins, you can use fuse wire to thread your beads, cutting it with old scissors and bending it with your fingers, protecting them if necessary.

If you have broken strands of jewellery beads, you can use these to create the drop-beaded designs on your candle collars.

3 ▶ PIERCING A HOLE IN THE COLLAR Use a darning needle to pierce a hole at the tip of each petal of the collar, resting the collar on a pad of kitchen paper.

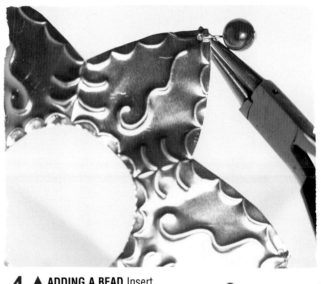

4 ▲ ADDING A BEAD Insert the hook of the bead into the hole and close the hook with the pliers, suspending the bead.

TEMPLATE COLLECTION – FOIL COLLARS

ideas from antiquity

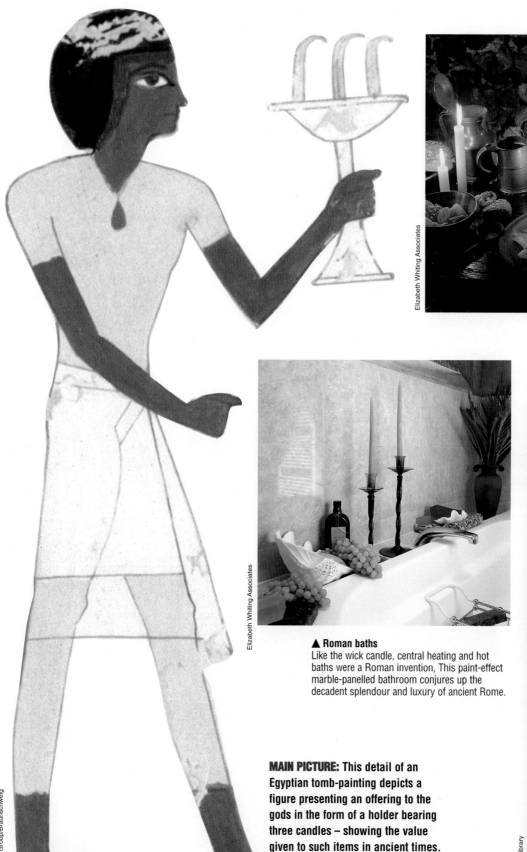

▼▲ Banishing darkness
Flickering candlelight, rough-hewn stone and rustic fare give a modern-day barbecue the feel of an ancient feast.

▲ Roman baths
Like the wick candle, central heating and hot baths were a Roman invention, This paint-effect marble-panelled bathroom conjures up the decadent splendour and luxury of ancient Rome.

MAIN PICTURE: This detail of an Egyptian tomb-painting depicts a figure presenting an offering to the gods in the form of a holder bearing three candles – showing the value given to such items in ancient times.

Candles have a very long history, but in the ancient world only the very wealthy could afford to use them to keep their homes brightly lit. In modern times, of course, that touch of luxury is within reach of everyone.

The Ancient Egyptians were probably the first people to use candles – examples made from beeswax have been found in tombs dating back to around 3000BC. These were made much like modern ones, although they used a reed for a wick, and were usually conical in shape.

Both the Old and New Testaments of the Bible mention candles, although what the writers had in mind would have borne little resemblance to the candles of today. They were made of reed and tallow (animal fat) and would have burnt more like flaming torches, with a very smoky flame. Certain types of garden flare are the closest modern equivalent.

As olive oil was plentiful in the Mediterranean, the Romans normally used oil lamps. When they conquered northern Europe, however, they adopted the tallow torches used by the local peoples.

Roman candles

The Romans are credited with inventing the first candles with a modern type of wick, in about 400AD. They were very expensive and were mainly kept for use in temples, although wealthier homes were also lit with them.

The word candle actually comes from the Latin *candere*, which means 'to flicker'. Early candles would indeed have flickered; the tallow from which they were made had a similar consistency to suet and burned poorly. And early wicks had a single braid that burned unevenly, which meant that they had to be regularly snuffed and trimmed.

Candles had a strong cultural importance in the ancient world. Roman history records the candlelit festival of purification celebrated during the month of February. This is believed to be the origin of the Christian festival of Candlemas, celebrated today on 2 February, when candles are blessed at the ceremony of the purification of the Virgin Mary, as well as having links with St Valentine's day festivities.

▲ **Torchlight procession**
Garden flares are the closest modern equivalent to the torches that our richer ancestors used to light their homes, although they are no longer considered safe for use indoors.

Roman candlestick
Made of terracotta, with a black, painted pattern, this candlestick looks remarkably modern in style. In fact, it is a replica of a Roman one found in the city of Pompeii, overwhelmed by the eruption of Mount Vesuvius in AD79.

reflective glow

INSET: A mantelpiece is the classic place to put both mirrors and candlesticks. Bear in mind that the candles will be seen against a background of what is reflected in the mirror. Here, this effect has been put to good use by matching the candles to the colour of the walls and choosing a contrasting colour for the flowers.

MAIN PICTURE: Double the impact of any candle by reflecting it in a mirror. And when the candles are burning in a stunning holder or wall-mounted sconce, it pays to make the most of it.

With an ever-increasing range of contemporary outdoor candles and lanterns, it's easy to use candlelight as the focus for a garden party or barbecue. There are even scented candles that will give a delicious perfume or keep insects at bay.

Terracotta container candles
There's a wide range of candles cast in containers designed to look good outdoors, or you can create your own from favourite flowerpots.

Garden torches
For a more powerful light, garden torches or flares have a stick that can be spiked into the earth, and a flame strong enough to resist a summer breeze.

Hooks, lanterns and flares
Put lights anywhere you want them with one of these, designed to spike into the earth or even in the lawn. Some metal hooks are designed for hanging garden lanterns, while others combine the stand with a shade or lantern on top. Bamboo flares look good in natural surroundings, and either have a lantern or an oil lamp on top.

summer breeze. The glass will give a wonderful sparkling light, and many lanterns look attractive in their own right. Examples on this page show just a few of the enormous variety available.

The second main option is to choose from the vast number of specially made outdoor candles. They range from glass or ceramic containers with a candle cast inside, to traditional garden flares that will stand in a lawn or flower bed. Outdoor candles with a natural scent, such as citrus or lavender will add a romantic touch to the evening, while citronella will keep flies at bay. Remember, you should never use an outdoor candle indoors, as its flame may be too fierce for a confined space.

Finally, you can use your imagination to create special outdoor candle holders. It could be something as simple as using old terracotta flowerpots in place of storm lanterns, or making a stunning (and windproof) arrangement such as the flower bowl shown on the left.

index

A
Adhesive	63
Adhesive Tape	65
Apple pomanders	55
Appliqué wax	49, 41, 42, 43
Appliqué wax motifs	40
Applying gilding	36
Aromatherapy candles	20

B
Base construction	21
Beeswax, rolled	16
Broke China mosaics	63

C
Candle Code (safety tips)	8
Carving candles	45
Carving dipped candles	44
Carving leaves	37
Chicken wire	79
Chinese paper lanterns	64
Cloves	55
Coloured layered cones	12
Coloured tealights	77
Coloured wax	15
Cone candle	25
Cones	12, 72
Cones, forming	73
Crushed ice	26
Cutting beeswax	18

D
Double boiler	12
Double mould	27
Dough, salt	57
Dye Pellets	21

F
Floating candles	61
Floristry foam	69
Foam, positioning candles in	71
Foam slicing & cutting	69-70
Frosted glass	61
Funnel for mould	15

G
Giftwrap candle cones	72
Gilded leaves	34
Glitter	47
Gluing tiles	61
Gold acrylic paint	63

H
Heart motifs	38-39
Heart template	39

I
Ice candles	39

J
Japanese symbol templates	43
Jewellery stones	61

L
Lavender aromatherapy candles	20

M
Masking tape	69
Melting wax	12
Metallic wax	35
Metail foil candle collars	80
Mosaic tiles	55
Mosaic votive holder	60
Motifs	41-43
Mould seal	25
Moulds, square	21
Moulds, cake tins	57

N
Negative stencilling	37

O
Orange pomanders	54
Oriental papers	65
Overfills	31
Pastel glow	76
Pebbles	79
Pillar candle	34, 39, 40, 42, 43
Pouring wax	30
Pre-cut appliqué wax sheet	41
Preparing the wick	30
Quadrant, drawing of	73

R
Radiant blooms	68
Rolling beeswax	17

S
Safety tips	8
Salt dough	57
Sanding holders	58
Scent, adding	23
Sealing mould	22
Seasonal flowers	69
Stars, sparkles & shapes	47
Stearin	22, 25, 29
Stearin melting	22

T
Tapered beeswax candles	18
Tealights	45, 54, 51, 76, 79
Tealights, glittered	46
Tealights, salt dough holders	59
Template, heart	39
Template, salt dough flower	59
Templates, Japanese symbols	43
Templates, using	38, 42
Tesserae mosaic squares	61
Thermometer	22, 26
Tiles, gluing	62
Trimming the wick	19

W
Warm hearts	38
Wax motifs	39
Wax pellets	25
Wick, adding in beeswax candle	17
Wick, keeping upright technique	14
Wick, preparation	13
Wick, trimming	19

READ MORE

Cerullo, Mary M. *Giant Squid: Searching for a Sea Monster*. Mankato, MN: Capstone Press, 2012.

Miller, Tori. *Octopuses and Squid*. Freaky Fish. New York: PowerKids Press, 2009.

Shea, Therese. *The Bizarre Life Cycle of Octopuses*. Strange Life Cycles. New York: Gareth Stevens Learning Library, 2012.

INDEX

A
Archie (the giant squid), 14–15
arms, 8–10, 18, 20–21

B
beaks, 10, 12, 22

D
dens (octopus), 22, 26–27, 29

E
eggs, 26–27
eyes, 8–9

F
food and eating, 10, 16, 22–23

G
giant Pacific octopuses, 18–29
giant squid, 6–17
gills, 20

H
Humboldt squid, 6–8
hunting by squid, 10

I
ink, 24–25
intelligence (in octopuses), 28–29

K
kraken, 5
Kubodera, Tsunemi, 16

M
mantles, 8–10, 20–21

O
octopuses, 18–29

R
reproduction (octopuses), 26–27

S
sizes of octopuses, 18
sizes of squid, 6, 8, 10, 15–16
skin (octopus), 18–20
sperm whales, 12–13, 16
squid, 6–17, 20
suckers, 10–13, 20–22

T
tentacles, 4–5, 8–9, 10, 14

oxygen (OK-sih-jen) The gas that humans and other animals need to breathe.

predators (PREH-duh-terz) Animals that hunt and kill other animals for food.

prey (PRAY) Animals that are hunted by other animals as food.

species (SPEE-sheez) One type of living thing. The members of a species look alike and can produce young together.

specimen (SPES-menz) A sample of something or an item to be scientifically studied.

tentacles (TEN-tih-kulz) Long, arm-like body parts.

texture (TEKS-chur) The way that the surface of something looks or feels. For example, smooth, rough, bumpy, or spiky are types of textures.

WEBSITES

Due to the changing nature of Internet links, PowerKids Press has developed an online list of websites related to the subject of this book. This site is updated regularly. Please use this link to access the list:

www.powerkidslinks.com/rlsm/squid/

GLOSSARY

aquariums (uh-KWAYR-ee-umz) Places where animals that live in water are kept for study and show.

chitin (KY-tin) A hard, natural material that forms the beaks of squid. It is the main material in the shells of insects, and the shells of crustaceans such as crabs and lobsters.

dissect (DY-sekt) To cut an animal or plant into pieces so that it can be scientifically studied.

folktales (FOHK-taylz) Stories passed down from generation to generation, usually by people telling the stories rather than writing them down. Folktales are often based on real-life events or explain something that happens in nature.

gills (GILZ) Body parts that underwater animals use for breathing. The gills take oxygen out of water and send it into an animal's body.

invertebrates (in-VER-teh-brets) Animals that have no backbone. This animal group includes ocean animals such as squid and octopuses, and land animals such as insects, which have hard outer shells instead of bony skeletons, as well as worms and snails.

Keepers in aquariums say that octopuses recognize the keepers' faces and will come out of their dens to beg for food.

THE SMARTEST SEA MONSTER

One thing that scientists have discovered over the years by studying octopuses in aquariums is that these animals are very smart!

Octopuses have been trained to find their way through mazes and tell the difference between squares and crosses.

Octopuses can also learn to carry out tasks such as unscrewing the lid of a jar to get to some food. An octopus named Ruby at the Biomes Marine Biology Center in Rhode Island was given a jar without a lid that contained a crab. Then, he was given a jar with a loose lid. As Ruby learned how to turn the lid, it was tightened. Ruby quickly learned to unscrew the lid to get to his crab treat!

SEA MONSTER MOTHERS

Female giant Pacific octopuses are very caring moms that actually give their lives to make sure their eggs hatch!

After mating, a female giant octopus finds a safe den and lays her eggs. She uses her saliva, or spit, to connect her eggs into strings. Then she sticks the strings to the roof of her den.

The female octopus watches over her eggs for six and a half months. She guards the eggs from predators such as crabs and sea stars. While she is waiting for her eggs to hatch, she doesn't hunt or eat. Once the baby octopuses hatch and swim away, the female octopus dies.

A female giant Pacific octopus lays around 57,000 eggs.

Giant Pacific octopus

Ink cloud

A predator may attack an octopus's inky cloud, thinking the cloud is the octopus itself. This gives the octopus time to get away.

ESCAPE TACTICS

A giant Pacific octopus is a large predator, but it still has predators of its own, such as sharks and other large fish.

Like most other octopus species, a giant Pacific octopus can release a cloud of thick black ink from its body if a predator gets too close. The ink spreads through the water so the predator's view of the octopus is blocked. Then the octopus makes its escape.

An octopus's ink also contains a substance that keeps the predator's sense of smell from working very well. This is important when avoiding sharks and other predators that use smell to find their food.

A dead shark, found on the seabed, is a feast for a hungry octopus.

FEEDING TIME

Giant octopuses catch and eat fish, crabs, and shellfish. They use their suckers to rip apart prey such as crabs.

Like squid, octopuses have hard beaks that they use for biting into their food. An octopus can also inject its prey with chemicals that poison the animal and soften its meat for eating.

Giant octopuses make dens among rocks. Inside a den, an octopus can eat and rest, safe from predators. After a meal, an octopus will dump trash, such as crab shells, outside the den's entrance. It uses water jets from its funnel to sweep its leftovers out of the den.

Sometimes fish visit the entrance to an octopus's den looking for a snack of octopus leftovers.

GIANT OCTOPUS PHYSICAL FACTS

Octopuses and squid are related sea creatures. Like squid, octopuses are invertebrates and have no bones in their bodies.

Also, like squid, octopuses have a combined head and body called a mantle. Octopuses have eight strong arms that they use for grabbing things. Each arm has two rows of suckers.

Octopuses breathe using body parts called **gills**. They suck water into their gills, take **oxygen** out of the water, and then blow the water out through a tube called a funnel.

Between an octopus's arms there is skin called a web. An octopus can suck water into its web and then quickly shoot the water out. This blast of water can propel the octopus along, fast!

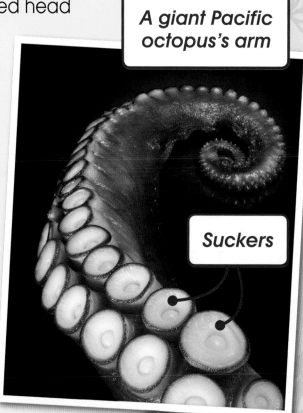

A giant Pacific octopus's arm

Suckers

A giant Pacific octopus

A giant Pacific octopus can even change its skin **texture** from smooth to bumpy to blend in with its surroundings and hide from enemies.

THE GIANT OCTOPUS

Along with huge squid, the world's oceans are home to another long-armed monster, the giant Pacific octopus.

There are around 300 different species of octopus. The giant Pacific octopus is the largest. Its arm span may stretch up to 30 feet (9 m)!

These real life sea monsters have a special skill. They can disguise themselves. To help it hide from predators, a giant octopus can change its skin color to blend in with its background. An octopus's skin contains thousands of tiny sacs filled with colored dye. Muscles around the sacs allow the octopus to enlarge the sacs or squeeze them tight to make different colors appear or disappear.

ENCOUNTER WITH A GIANT

In 2004, Japanese scientist Tsunemi Kubodera and his team became the first people to capture photos of a live giant squid in its natural habitat.

At a depth of nearly half a mile (0.9 km) beneath the ocean, the scientists used a long fishing line baited with food that is attractive to giant squid. As a huge squid attacked the bait, underwater cameras took more than 500 photos of the creature. The team estimated that the giant squid measured about 25 feet (7.6 m) long.

The Japanese team had decided to follow sperm whales. They hoped the whales would lead them to places where the team could find the mysterious giant squid. Their plan worked!

In 2006, Tsunemi Kubodera and his team achieved another first. They became the first people to catch and film a live giant squid.

Archie, the giant squid

Archie measures just over 28 feet (8.6 m) long. Scientists believe Archie may actually be female, but more studies are needed to be sure.

MEET ARCHIE

Fishing boats sometimes accidentally catch giant squid in their nets. In 2004, a giant squid was caught near the Falkland Islands in the South Atlantic Ocean.

The enormous **specimen**, nicknamed Archie, was frozen and transported to the Natural History Museum in London. Scientists at the museum decided not to **dissect** Archie, but to preserve the squid whole.

It took three days for Archie to defrost. Then scientists unraveled the squid's tentacles and measured the animal. Finally, Archie was put on display in a super-long tank of water. The water contains salt and a chemical called formalin, which will preserve the squid's body.

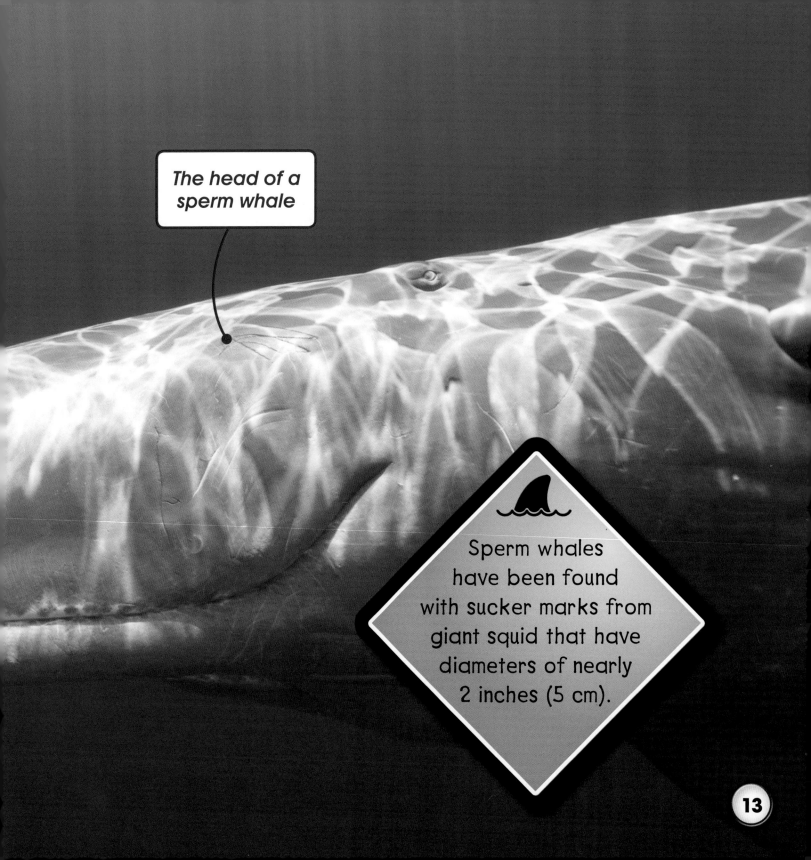

The head of a sperm whale

Sperm whales have been found with sucker marks from giant squid that have diameters of nearly 2 inches (5 cm).

MEGA BATTLES

Giant squid are among the largest predators that live in the world's oceans. These huge predators have their own enormous enemies, though.

A giant squid's main predators are sperm whales. Weighing many tons (t) and growing longer than school buses, sperm whales travel the world's oceans hunting giant squid. Their prey, however, do not give up without a fight! Many sperm whales carry the scars, including large sucker marks, of their mega battles with giant squid.

Sperm whales have helped scientists find out about giant squid. The whales cannot digest the squids' hard beaks. When whales are washed up on beaches, and scientists examine their stomach contents, the scientists often find giant squid beaks that can then be studied.

A close-up of scars made by giant squid

Suckers

Circles of sharp, teeth-like chitin

Inside a squid's suckers are rings of a hard substance called **chitin** that sink into its victim's flesh like sharp teeth.

A GIANT HUNTER

Scientists believe giant squid only live for about five years. In order to get so large, these animals have to grow fast. To grow fast, giant squid must catch and eat vast quantities of fish and other squid.

A giant squid catches its **prey** with body parts called clubs on the ends of its tentacles. Suckers on the animal's clubs, tentacles, and arms attach to the prey and hold on with powerful suction.

Once its prey is trapped, the squid pulls its meal toward its mantle. The squid's arms help hold the struggling creature and drag it into the squid's sharp, parrot-like beak.

A squid's beak is inside the mantle. The sharp beak easily breaks the prey animal into small pieces.

A giant squid's beak

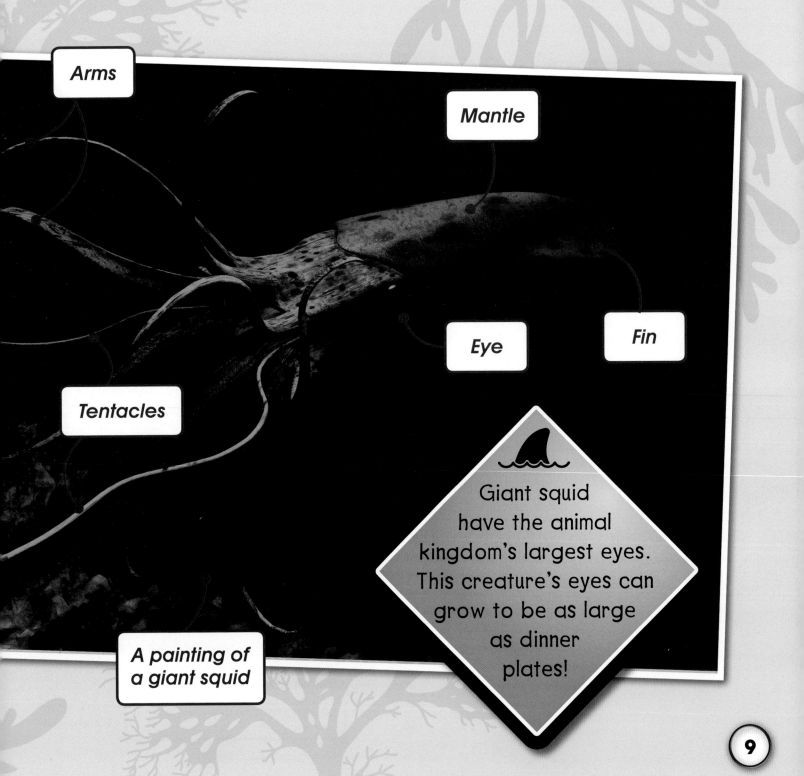

A painting of a giant squid

Giant squid have the animal kingdom's largest eyes. This creature's eyes can grow to be as large as dinner plates!

GIANT SQUID PHYSICAL FACTS

Squid are invertebrates, which means they are animals without a backbone.

A giant squid has a combined head and body called a mantle. At one end of the mantle are the animal's fins. At the other end are eight powerful arms and two extremely long tentacles.

A giant squid's arms can grow to be over 9 feet (3 m) long. Some especially large giant squid may have tentacles that are up to 40 feet (12 m) long.

The giant squid's record-breaking total-length measurement includes its mantle and its tentacles. These huge animals can reach weights of 880 to 2,000 pounds (400–900 kg).

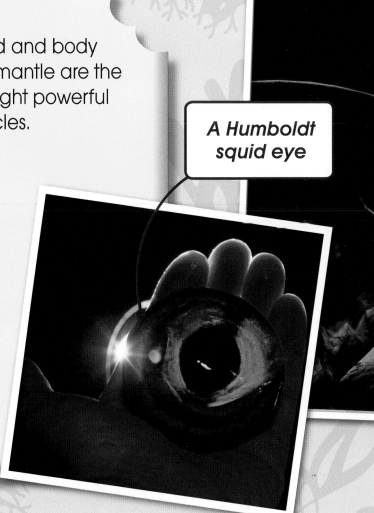

A Humboldt squid eye

Scientists have discovered most of what they know about giant squid by studying dead giant squid that wash up on beaches.

REAL LIFE SEA MONSTERS

Stories of giant, tentacled sea monsters probably came from sightings of real life animals such as octopuses and squid.

The animal pictured here is a large **species** of squid called the Humboldt, or jumbo, squid. These ocean hunters grow to around 6.5 feet (2 m) in length. Many sea monster tales may have begun with sightings of the truly immense giant squid. There have been reports of these squid growing to lengths of 60 feet (18 m)! An average length for the giant squid, however, is 40 feet (12 m).

Humboldt squid have been photographed and studied in the world's oceans. No one had ever photographed a live giant squid in its natural habitat, however, until 2004!

Folktales from Norway told of the kraken. This enormous sea monster had horrifying, grasping arms, soccer ball-sized eyes, and powerful jaws.

A SEA MONSTER ATTACKS!

A large ship sails on a calm sea. Suddenly, a forest of enormous, arm-like tentacles breaks the surface of the water.

A massive sea creature emerges from the ocean and begins to wrap its tentacles around the ship. The sailors hack at the beast with axes, but they are doomed to a watery death, as the hungry monster pulls their ship beneath the water.

For centuries, people imagined the world's oceans were home to giant sea monsters. Stories of monsters attacking ships were passed down from generation to generation. Today, we know these horrifying creatures do not exist. So how did the terrifying stories of killer sea monsters get started?

CONTENTS

A Sea Monster Attacks! 4
Real Life Sea Monsters 6
Giant Squid Physical Facts 8
A Giant Hunter 10
Mega Battles 12
Meet Archie 14
Encounter with a Giant 16
The Giant Octopus 18
Giant Octopus Physical Facts 20
Feeding Time 22
Escape Tactics 24
Sea Monster Mothers 26
The Smartest Sea Monster 28
Glossary 30
Websites 31
Read More 32
Index 32

Published in 2014 by The Rosen Publishing Group, Inc.
29 East 21st Street, New York, NY 10010

Copyright © 2014 by The Rosen Publishing Group, Inc.

All rights reserved. No part of this book may be reproduced in any form without permission in writing from the publisher, except by a reviewer.

Produced for Rosen by Ruby Tuesday Books Ltd
Editor for Ruby Tuesday Books Ltd: Mark J. Sachner
US Editor: Joshua Shadowens
Designer: Emma Randall

Photo Credits:
Cover, 1, 4–5, 18–19, 20–21, 28–29 © Shutterstock; 6–7, 11, 12–13, 22–23, 27 © FLPA; 8–9, 24–25 © Superstock; 10 © Science Photo Library; 14–15 © Getty Images; 16–17 © Press Association.

Library of Congress Cataloging-in-Publication Data

Owen, Ruth, 1967–
　Giant squid and octopuses / by Ruth Owen.
　　pages cm. — (Real life sea monsters)
Includes index.
ISBN 978-1-4777-6261-5 (library) — ISBN 978-1-4777-6262-2 (pbk.) — ISBN 978-1-4777-6263-9 (6-pack)
1. Octopuses—Juvenile literature. 2. Squids—Juvenile literature. I. Title.
QL430.3.O2O937 2014
594'.56—dc23

2013029147

Manufactured in the United States of America

CPSIA Compliance Information: Batch #W14PK7: For Further Information contact: Rosen Publishing, New York, New York at 1-800-237-9932

REAL LIFE SEA MONSTERS

Giant Squid and Octopuses

by Ruth Owen

PowerKiDS press

New York

JUL 15 2014

Sayville Library
88 Greene Avenue
Sayville, NY 11782